Schroff Development Corporation
P.O. Box 1334
Mission, KS 66222
(913) 262-2664
www.schroff.com

Raudebaugh, Robert A.

Visualization, Sketching and Freehand Drawing for Engineering Design/Robert A. Raudebaugh

ISBN 1-58503-003-1

Printed and bound in the United States of America

Preface

The profession of engineering has always held that visualization is one of the most important skills in engineering and many other areas of design. The problem is, visualization is not a well understood skill and even less is known about the process of developing it. However, there is a growing body of evidence indicating that the exercises and processes set forth in this text are the key to developing these skills. Even though it may not be well understood in engineering, this developmental process has been known by artists for ages. According to William Hogarth, an 18th century artist:

> He who understands the acquisition of perfect ideas of the distances, bearings, and oppositions of several material points and lines in even the most irregular figures, will gradually arrive at the knack of recalling them into his mind when the objects themselves are not before him - and will be of infinite service to those who invent and draw from fancy, as well as to enable those to be more correct who draw from live.

The concepts, objectives and activities contained in this volume have been developed after several years of careful study and experience in the classroom, two research projects and extensive review of the literature. The primary goal of the studies is to extend the perceptual skills of the student, based on the assumption that this will result in improved visualization skills, which will in turn serve the student in the design process. The nature and the sequence of the studies have been well documented. Experience has shown that if students follow this sequence and do the exercises conscientiously, they will learn to draw well.

The illustrations in this text were all done by the author with two exceptions of student work. There is no special "artistic talent" involved. The skill level demonstrated, which is at least competent, was developed following the processes outlined in this text. The background of the author is somewhat typical except for training and experience in drafting and engineering graphics, but included no art education or training. Many of the author's students, beginning with even less skill, have far exceeded those of the author. The point is, the author is living proof that the process does, in fact, work.

Acknowledgements

No idea belongs solely to one individual. I would like to thank the Engineering Graphics Faculty at Arizona State University who worked collectively on these ideas from 1984 to 1988. I would like to particularly thank Professor John Matson without whose assistance this project would never have even begun and also Professor Del Bowers, whose ideas kept feeding the process for those formative years. At Western Washington University, I would like to thank Professor Kathleen Kitto who believed in the idea and helped keep it alive and Professor Jeffrey Newcomer who contributed the chapter on design and whose editing assistance was a valuable asset.

Dedication

I dedicate this Work to my loving wife Linda without whose belief and assistance the project would never have been completed, and to my two sons, Erich and Kevin whose editing efforts and support proved invaluable. Thank you!

Table of Contents

References (for additional reading)

Betti, Claudia and Teel Sale. Drawing: A Contemporary Approach. Second Edition. New York: Holt, Rinehart and Winston, 1986.

Dodson, Bert. Keys to Drawing, Cincinnati, Ohio: North Light Books, 1985.

Edwards, Betty. Drawing on the Right Side of the Brain. Los Angeles, J.P. Tarcher, Inc. 1979.

Frank, Fredrick. The Zen of Seeing. New York: Vintage Books, 1973.

Frank, Fredrick. Zen Seeing, Zen Drawing, Meditation in Action. New York, Bantam Books, 1993.

Hale, Jonathan. The Old Way of Seeing. Boston: Houghton Mifflin Co., 1994.

Lockard, William Kirby. Design Drawing. Tucson, Arizona: Pepper Publishing, 1982.

McKim, Robert H. Experiences in Visual Thinking. Second Edition. Montery, California: Brooks/Cole Publishing Co., 1980.

Nicolaides, Kimon. The Natural Way to Draw. Boston: Houghton Mifflin Co., 1941.

Chapter 1 - Introduction

Why?

Freehand drawing is not something often associated with the process of engineering, so it is legitimate to ask the question: Why teach it to engineering students? Numerous technological advances since WWII have reshaped the way in which engineers work. One of the most significant advances in recent years has been the desktop microcomputer. In order to incorporate these changes into engineering curriculum without loosing any essential skills, numerous studies have been conducted in the past 10-15 years by engineering educators for the purpose of ascertaining just what it is that engineers do and the kind of education they need. Several such studies have focused on the graphics skills engineers need, and as a result of these efforts, a new national curriculum for engineering graphics has emerged. This curriculum, developed at the University of Texas by Ronald Barr and Devorac Juricic, identified three levels of drawings (graphics) associated with the engineering design process: 1) ideation drawings, 2) communication or development drawings, and 3) documentation drawings.

Ideation drawings are those drawings which are associated with the process of generating ideas in the early stages of design. The purpose is to record ideas in rapid succession so that the designer may explore a wide variety of possibilities in a short time span. The more ideas, the better the ultimate design. Words are usually too awkward to describe an idea of a physical object, and words can inhibit visual thinking and the creative process. The idea sketch need not be elaborate; just enough so that the designer may recall the idea at a later time after the ability to generate new ideas has been exhausted and it is time to evaluate them to determine which of these, if any, have promise.

Communication or development drawings represent the next step in the process where a very basic sketch is elaborated upon as the idea is explored further. The communication sketch may be extended further as details are worked out for the most promising ideas. Once a decision is made to try an idea as a solution for a problem, the sketch may need to be developed further into a drawing through which the idea may be fully communicated to others. This is carried to its fullest extent in industrial design where the communication drawing is developed to photographic quality for potential client review or possible advertising campaigns before the product is fully developed and prototypes are available. The amount of detail required depends on the design stage, the skill of the designer, the time available, and the visualization skills of those to whom the designer is communicating. However, it is important to realize that the ultimate responsibility for communication lies with the designer. The greater the detail and drawing fidelity required, the greater the skill requirements are for the designer. In engineering, the communication drawings may take the form of technical sketches, which are necessary to include the important details of the design.

Documentation drawings represent the final stage in the engineering design process. These drawings incorporate all of the detail required to manufacture the part to the specifications and quality standards of the designer, and have traditionally been referred to as "engineering drawings". In the past, manual drafting techniques were used to develop engineering drawings and draftsmen were specifically trained to produce them. Today, most engineering drawings are developed with Computer Aided Design Drafting software on either microcomputers or desktop workstations. As computers and CADD software advance, the nature of how these documents are created is changing. At first, microcomputers were used in a manner similar to manual drafting tools to produce 2-diminensional multi-view drawings using the system of orthographic projection and descriptive geometry developed in the mid-18th century. As microcomputers grew in power, software was developed which was capable of producing 3-dimensional models of parts and assemblies. Early computers and software capable of producing 3-D models were very expensive; today however, powerful 3-D systems are available at affordable prices for almost every level of design function. As both micro and workstation computers and software grow even more powerful, integrated systems on which designers can create the model of a part, analyze it for necessary engineering properties, and then create the program for manufacturing the part on computer controlled machines are becoming increasingly common. Corporations using these systems are finding that engineers tend to want to do all of the design work on the computer. However, they are discovering that when engineers do not begin with sketches, they soon loose their ability to think visually.

One of the most essential graphic skills, known for some time and reemphasized in recent studies, is that of visualization. How does it fit into the different types of drawings? There is strong evidence to indicate that it is essential to all forms of engineering drawing. The advancements in computers which now allow engineers to create 3-dimensional design models directly, necessitate that engineers have the ability to think 3-dimensionally, and visualization plays a major part in this process. Where do visualization skills come from? This is not well understood, but both the latest research and classroom experience indicate that perhaps the most powerful tool for developing visualization skills is freehand drawing.

How?

Freehand sketching is not a skill commonly found among engineering students; in fact it is not a skill common among the general population. Most people educated in the United States and in most western societies lack this skill, and the majority believe that the development of the skill is beyond their capabilities. Freehand sketching or drawing skills have been relegated to those few unusual persons thought to possess that magical "artistic" ability. Science has not discovered much about where drawing skill comes from, but we do know that those who have it are relatively few in number. The important question raised by this perception (not necessarily fact) is: if freehand drawing/sketching is such an important skill, and it is so rare, can it be learned by engineering students? If so, how?

Experts in the field such as Rudolph Arnheim and Howard Gardner tell us that drawing is a natural skill and that everybody can develop this skill to a competent (if not artistic) level. If this is so, then why do so few people in our society possess it and why are so few of them engineers or engineering students? To answer this question, we must take a look at our educational system and at the cognitive values in our society.

Those who control how we educate are among the most convinced believers that drawing or artistic skill is possessed by a relative few and that you are either born with it or not. Art education in America is not about developing drawing or visualization skill, or doing art, but more about exploring and learning about art. If we examine closely those topics that are favored most by the educational system and supposedly are a reflection of society, the famous three "R's" immediately come to mind; reading, 'riting and 'rithmetic. Throw in social studies and a little science, and you have the "core" curriculum of every school system in America. Where is drawing? It has some favor in pre-school and kindergarten classes, but primarily as entertainment or to aid developing motor and color recognition skills. It is highly unlikely that very many early childhood teachers understand the need for, or how to develop visual and drawing skills. In the early grades art is not well understood, and is frequently relegated to one or two periods a week, taught by an art specialist who must deal with all of the classes in a school and not infrequently, several schools, if such a person exists at all. In-class activities again include those used to help children learn colors or to improve motor skills and neatness attitudes by keeping within the lines of mass-produced outlines.

If we examine the average middle class home in America, we will see reflected a similar set of learning values. Much has been written about the importance of reading in school and how important it is in the home if a child is to have an easy time of it. Children are also frequently aided in learning numbers and counting in the home. What about art in the home? Few families place a high value on free and expressive art activities. Most home "art" activities consist of coloring books, where children place "appropriate colors in appropriate places in appropriate pictures" and, of course "must stay within the lines". Free art exercises may have been met with some displeasure on the part of parents, particularly if those efforts involved an inappropriate drawing substance such as lipstick on an inappropriate medium such as a floor or wall. There are not many enlightened parents who will encourage the activity and aid the child in selecting a more appropriate medium. Later in middle school and high school there are elective courses where an "academic" approach to art - or what Betty Edwards referred to in Drawing on the Right Side of the Brain as "mucking about with art"- is taken. Art classes might also provide a place where those students who exhibit the rare "artistic talent" which developed in spite of the system, can go to practice their skill. All too frequently these students are thwarted by teachers who do not understand the cognitive processes associated with art or with drawing.

Howard Gardner has recently published a work on "multiple intelligence" which has become popular in the education world. Many educators are currently making a strong case for the need to incorporate or somehow make allowances for multiple intelligence in our educational system. It is possible that recognition of the various ways that people process information could have a significant and positive impact on art education. If this is like many of the other "great discoveries" which have flooded our educational system, it will be a long time until the knowledge of multiple intelligence has any effect on what is done in school, particularly in the area of visual education.

It is more likely that the answer to how we can learn to draw or to teach drawing lies in the work of Roger Sperry, who in the 1950's discovered that the cerebral hemispheres of the brain process information quite differently and independently. Sperry's work and that of his colleagues and numerous others is popularly known as "split brain theory" or "right brain - left brain theory". This theory states that most verbal functions occur in the left hemisphere of the brain while most visual functions occur in the right hemisphere. This is not a new theory. The French physician Paul Broca, noticed as early as the mid 1800's, that patients with damage to the left side of the head and brain were aphasic (could not speak). He concluded then that the speech function must be located in the left hemisphere of the brain. His conclusions were not accepted by the medical profession until after his death and now the egg-sized area of the left hemisphere of the brain which is known to contain the speech function is also known as Broca's area.

What does all of this have to do with learning to draw? Simply, our educational system and society neither value nor understand activities which develop the cognitive processes now known to be associated with the right hemisphere of the brain. We do not train children or adults visually. If we examine how people naturally learn to draw, those who have studied the skill carefully believe that drawing is a teachable, learnable skill, much the same way in which language is a teachable, learnable skill. Howard Gardner identifies different levels of natural visual or artistic development. At first, around the age of 2, the child discovers sideways motion while holding a drawing implement. As children develop the necessary psychomotor skills, they begin to include circular motions as well. Sometime around the age of 5, the child begins to recognize certain patterns in the resulting lines, and begins to assign names to these patterns, such as a tree or a bunny or house. These continue to evolve until the child develops a set of symbols for most of the common objects in his or her life. In our society, these symbols are highly consistent, and represent mostly what we know intellectually about the object rather than reflecting what is seen in individual objects. Children eventually make the discovery that their work does not represent the "real world" very accurately during early adolescence. Then comes the age of experimentation, when the adolescent child deals with this discomfort with rather bizarre treatments of common objects. Finally, the desire to achieve realism becomes too strong and either the adolescent enters the age of realism, if properly assisted, or without proper assistance at this critical stage begins to believe that they lack that certain "artistic talent" and ceases all attempts at drawing. Unfortunately the latter of the two possibilities in the most common, primarily because drawing is not particularly valued in our society and very few art teachers have the skill and understanding necessary to provide the assistance required to help students learn to draw well. Educators often explain this inability to draw, by saying that person just does not have "artistic talent" necessary. In spite of the system, some people do learn to draw.

Why? Why does an 8 year old learn to play classical music on the violin or piano? Why is one individual able to accomplish astounding feats of athleticism at an early age; why does one gain the skill to write a best seller? How can an adult pick up a guitar for the first time and play it beautifully without lessons? Why is a person able to play a tune on the piano after just hearing it? Why is one person able to hold an audience spellbound while an equally intelligent and knowledgeable person bores them to tears? All of these conditions exist and we don't have complete answers as to why. The fact is that they represent the extreme and not the common condition. Since we do not value it strongly and since we do not understand it well, drawing has been wrongfully attributed to the extreme or uncommon set of conditions.

The answer which may help us deal with all of this may be found in the research done on split brain theory and the different ways that the two cerebral hemispheres process information. The problem is that when as adults we attempt to draw, we do so based on what we know about something, not what we see in that object. We have been trained verbally, and have stored a wealth of knowledge about the world around us, which in many endeavors serves us well. Drawing is, unfortunately, not one of them. When trying to learn to draw, we typically look at an object for a few seconds, and then focus on the paper or other medium and attempt to capture it from short term memory. The brain is capable of holding an image only for a very short time. When the time we look at the paper exceeds the image retention time, we remember stored symbols associated with the object. We then begin to rely on what we know; that information about the object that has been relegated to long term memory. Thus the lines on the paper begin to represent the symbolic rather than realistic. The answer lies in learning to see. Learning to see in this sense does not involve visual acuity or visual perception, but learning to focus on the visual information in the object such as line, contour, space, angle, proportion, shape, and other spatial relationships, rather than stored facts and symbols.

Kimond Nikolaides (The Natural Way to Draw) discovered the techniques for accomplishing this type of learning to see as a teacher in the New York Institute of Art in the 1920's and 30's. Betty Edwards (Drawing on the Right Side of the Brain) made the association between artists' seeing and hemispheric cognitive processing from recent discoveries on the workings of the brain. Both believed that anyone could learn to draw, drawing being a natural experience, if they could learn to use the visual skills associated with the right hemisphere of the brain, or if they could learn to process information in a purely visual fashion.

Before you attempt to train the brain to process information visually, there are some experiences that are important to have first. Before you can be successful, you must convince your psyche that the above claim is true.

Chapter 2 - Experiments in Visual processing

Flash Drawing

In order to convince yourself that there is a visual way to process information which is more effective than how we might normally approach a drawing task, complete the following two visual experiments. The first we will call a "flash drawing". Find a photograph in a magazine or perhaps from the family photo album of an object with which you are fairly familiar, possibly something that you may have attempted to draw on a previous occasion, or you may use the image found on page 11 of this text. Use the following directions for the experiment. Read through all of the directions before you begin.

1. Get a blank sheet of paper and place it on a hard surface in front of you where you can access it comfortably. Use a soft lead pencil; a #2 office pencil will do fine.

2. Have a clock handy so that you can time yourself for this experiment. Allow a maximum of 15 minutes.

3. Turn the picture upside down and leave it that way for the duration of the experiment.

4. Place the picture before you and stare at intently to the count of ten. Try to memorize all of the features in the picture.

5. As quickly as you can, (do not go over the allotted 15 minutes) and using the paper and pencil, make a drawing of what was contained in the picture as accurately as possible. Try to remember all of the details by recalling the names of objects contained in the picture. Remember that

the object of this experiment is to produce as accurate of a copy as possible.

6. Complete the experiment before reading on.

When you have completed the task, turn the picture over and compare it to the copy that you just attempted to make. How accurate is the copy? Now think back over the experience. Did you spend any time looking at the clock? If so, did this cause any degree of anxiety? Were you able to recall any names of the objects contained in the picture as you were attempting to draw them? Did you have difficulty getting them "right"? If you answered 'yes' to any of the questions above, you are in the normal range for a non-artist. The instructions were designed to activate the verbal or left hemisphere of the brain. This should have created conflict between the two hemispheres as the right one was trying to focus on what was essentially a visual task, but kept getting interrupted by recalling names, keeping track of time, or worrying about how closely it would resemble the original, all left hemisphere tasks.

Flash Drawing

Upside Down Drawing

Now let us try another experiment, but this time using a visual approach. Consider the upside down drawing of Einstein on the following page. Do not turn your book around; it is that way for a good reason. Get another piece of paper and use the following set of instructions for this experiment. Read through all of the instructions before you begin.

1. Place the paper before you so that you can access it comfortably. Place the book so that you can view the figure without holding on to the book.

2. Make sure that you have plenty of time for this experiment (set aside at least 45 minutes), *then forget about time.*

3. Begin somewhere along the top of the frame. For a few moments, just look at the picture. Notice the lines that make up what is there. How long are they? How do they relate to the other lines around them: angle, shape, proportion? Where on the page do they begin and how far do they go? What is the angle that they make with the edge of the page? What are the shapes that the lines enclose?

4. Focus on one line. Notice the spatial characteristics of *that* line. Where does it begin, where does it end, at what angle? Finally, on your paper, carefully record that line.

5. Select an adjacent line. Notice the same spatial characteristics of the next line. Also notice the relationship to the line just recorded, then record the second line.

6. Select the next line and continue the process for each successive line. Work as though you are putting together a puzzle using all previous lines to establish the spatial data for the next line.

7. Should a name of some feature in the picture occur to you as you look at a set of lines, put it out of your head, or pause, focusing on visual

information until the name goes away. If the name persists, go on other lines and come back when you are able to suppress the name.

8. Continue in this manner until you have recorded all of the lines in the figure. When you are done, turn both the book and your drawing right side up and reflect on the experience.

Are you surprised at how well the drawing turned out for you? Did the names come and were you able to repress them? Did you loose track of time while working, and did you feel rested and refreshed at the end of the experiment? If you can answer yes to most of these questions, you have experienced what artists refer to as the right mode shift. You were processing in a basically visual mode, which is the way you need to process visual information if you wish to learn to draw.

What was happening in the two different experiences and how can we use this to help learn to draw? First of all, it is important to understand something about the way the brain processes information. As you may know, the brain is divided into parts; the brain stem, the inner brain, and the cerebral cortex. It is the third or outer layer that contains all of the higher order processes including thought, memory, and information processing. Experiments over the last 50 years have confirmed that the two hemispheres of the cerebral cortex process information differently and independently. The left hemisphere processes verbal and written information. Since that type of information is also symbolic, abstract, and linear, these processing characteristics have also been attributed to the left hemisphere of the brain. Other tasks which have these characteristics, including those dealing with mathematics, memorizing information, facts, or tasks which are essentially language based, have all been attributed to the left hemisphere of the brain.

What goes on in the right hemisphere is less easily determined, but it has been pretty well established that visual tasks take place in the right hemisphere. The right brain also processes concretely in visual images.

For example, the left hemisphere would be involved in analyzing a rock, describing its shape, color, mass, size, and mineral content. The right hemisphere of the brain would be involved in the mental image of the rock, possibly making mental/visual comparisons with other rocks or "seeing" analogies or relationships that the rock brings to mind concerning other objects or other rocks. The table on page 15 gives a brief summary of the cognitive processing characteristics of the right and left hemispheres of the brain.

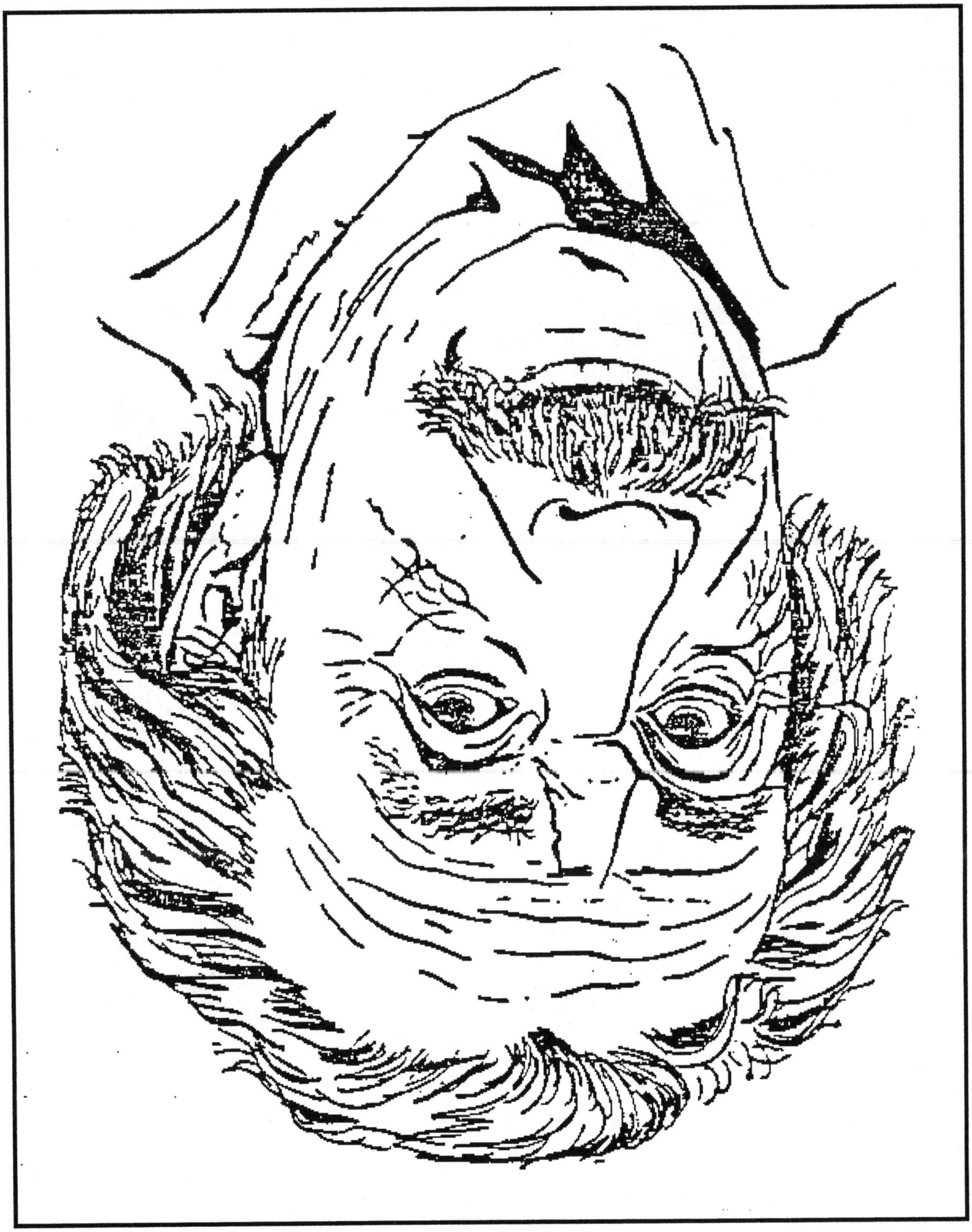

Upside-Down Drawing

LEFT Mode	**RIGHT** Mode
Verbal - dealing with words, both written and orally.	Visual - dealing with images
Linear - Sequential	Concrete
Temporal	Analogies
Abstract	Loose track of time
Numerical	

There are several things to keep in mind when thinking about hemispheric laterality. First of all, there is no clear dichotomy of cognitive processes. The human brain is far too complex to make that simple of a comparison. Since humans are complex and different in many ways, the degree of lateralization naturally differs as well. The most highly lateralized group is right-handed males. Women are less so, left-handed males even less so, and there is some evidence to indicate that left-handed females whose mothers were also left-handed may have their cognitive functions totally reversed. The occurrence is so rare that this is difficult to verify. The point is that all of us have certain cognitive functions lateralized to some degree in one or the other of the two hemispheres of the brain. So what does this mean and what does all of this have to do with learning to draw?

Our educational system concentrates almost exclusively on developing those cognitive skills associated with left hemisphere processing; reading, writing and math. By far, the most important skill for success in school is that of reading, as reading (along with lecture, also a left mode process) is the most often used way of obtaining "knowledge". Those activities which might develop right-mode processing skills such as learning to draw from observation, making things from drawings or free and expressive art activities, are almost completely missing from the elementary grades. They are given only minor roles in the secondary grades, and are frequently reserved for those students considered less academically capable or those considered to have that special "artistic talent". This situation in American education led one researcher in this field to the conclusion that our educational system K through graduate school virtually destroys the right hemisphere of the brain. At the very least, it has made the task of learning to draw a very daunting one for most of us.

In all of this, there is hope. The trick is to regain some of the visual processing skills that have been neglected, and to learn how to quiet the verbal left hemisphere when attempting to process visually. The previous two exercises are an indication of the problems which can occur when the left mode interferes, and the success that can be had when the left mode is quiet and the right mode takes over. Essentially, learning to draw successfully means learning how to process information visually; learning to see as an artist sees, and at the same time, learning to quiet the left mode. The exercises in chapters 3 through 8 are specifically designed for this purpose.

Chapter 3 - Getting Started

Do You Have Sufficient Dexterity?

Many people are convinced they cannot learn to draw because they lack the necessary dexterity, or ability to make the hand and fingers do the proper things with the pencil. They have tried, but the pencil just will not go where it is supposed to. By now, you may suspect that possibly some sort of conflict between the hemispheres or cognitive functions of the brain may be the real culprit. In order to find out, you need to take the artist's dexterity text. Use the following instructions.

1. Take out a clean piece of paper and place it long-ways in front of you, or, horizontally. This is referred to as landscape orientation.

2. Using a pencil or any other handy medium, sign your name, in cursive and as large as possible. Try to fill the whole page.

3. If successful, you have just produced a drawing, and at the same time proven that you have sufficient dexterity to learn to draw.

This may seem somewhat simplistic, but the point is that dexterity is not the issue, seeing is.

Author's Signature

Your signature is a form of line drawing also known as a contour drawing. The particular formation of the contours has specific and symbolic meaning to anyone able to read your writing and able to understand written language. A popular myth about art and drawing, and there are many, is that knowing how to draw is being able to make the hand and pencil go exactly where you want it to. Although controlling the pencil is important (this is a skill that can be developed) the real problem is seeing; that is, developing the ability to see the way an artist sees, in a purely visual format, using right mode cognitive processing. Being able to sign your name is an indication that you posses sufficient manual dexterity to learn to draw well.

Overcoming the Blank Page

Even though you have just proven that you have the appropriate dexterity, getting started can still be quite intimidating. It need not be. One of the biggest obstacles to learning to draw, other than learning to see, is overcoming fear, particularly fear of failure or not being able to get it "right". As was stated earlier, one of the myths associated with drawing is that drawing is being able to control where the pencil goes, and although we now know that is not totally true, there is still an element of truth in it. Just as with any motor skill, learning to draw takes some practice and some time to develop, particularly the drawing of straight lines. Probably no one is able to draw straight lines on the first try. Those who have developed this skill just did not give up on the first try, so let us start with some straight lines.

All you need is a piece of white paper; it does not really need to be white, but white seems to work best. Any kind and size will do, but the best results will be had with a decent bond drawing paper about 9x12 inches. Any pencil will also do. For beginners, a pencil is better than a pen and a soft lead (No. 2) office pencil will do as well as any.

Begin with horizontal lines. Fill the first page with them, varying the distance between. Do not be concerned with whether or not they are straight at first. That will come with practice. Try to make them straight, but do not get concerned if they are not. Now try a second sheet, but this time try varying the line width by rotating and tilting the pencil point. Play with different patterns.

Next, fill several pages with vertical lines using the same technique. Again, try making them straight but do not get concerned if they are not. Just relax and enjoy the experience. Try for some creative patterns by varying the line width and spacing between lines. When you are tired of this, try the same thing with diagonal lines.

Now try a page with circles. Vary the size and allow some to overlap. Work at getting round circles, but it is not important that they be perfect. They will improve with practice. Try also for creative patterns. Being creative makes it more interesting, both to do and to look at later. Next go on to ellipses and try to accomplish the same things as with circles. Try a page of combinations of circles, ellipses and circles then lines. Most importantly, be creative and enjoy the experience.

There are techniques for drawing straight lines. It is not important to master the technique, but to use it as a tool in helping to get straighter lines in the beginning. Later, you will be able to make straight lines without using any aids. For now, the procedure is as follows. Place two points on the paper. Make them fairly close at first. Decide which direction you want to draw the line and place your pencil tip on the first point. Then fix your eye on the second point. While holding your eye fixed on the second point, move the pencil toward it. Fill the page with points and practice this with variations, making the points at different distances and the lines at different angles. You will find that "taking" a line to a point you have fixed in your vision is akin to keeping your eye on the ball when trying to catch or hit it, or keeping your eye on the target while throwing something at it. It works!

Lines make up the essence of all shape and form in drawings. Lines in objects are formed by contours and edges. A contour is where the an object meets space, as in the earth's horizon. To see contours of objects, we must ignore the background and focus exclusively on the object. Edges are formed when two surfaces meet. This is easiest to see when the surfaces are flat. In that case, we refer to the edge as a corner, and if we are drawing such an object, we represent the corner or edge with a line. However, the surfaces meeting to form an edge do not have to be flat. If we look at the curved surfaces on the inside of our hand for example, we will see many edges formed where two surfaces meet. We are unlikely to find any flat surfaces on our hand. Such is the case with most objects in nature. There are not many flat surfaces, or "corners" but there are many instances where curved surfaces meet to form edges or where objects meet with space to form contours. For the sake of simplicity, we will refer to all such meetings as contours. In drawings, these contours are represented with lines.

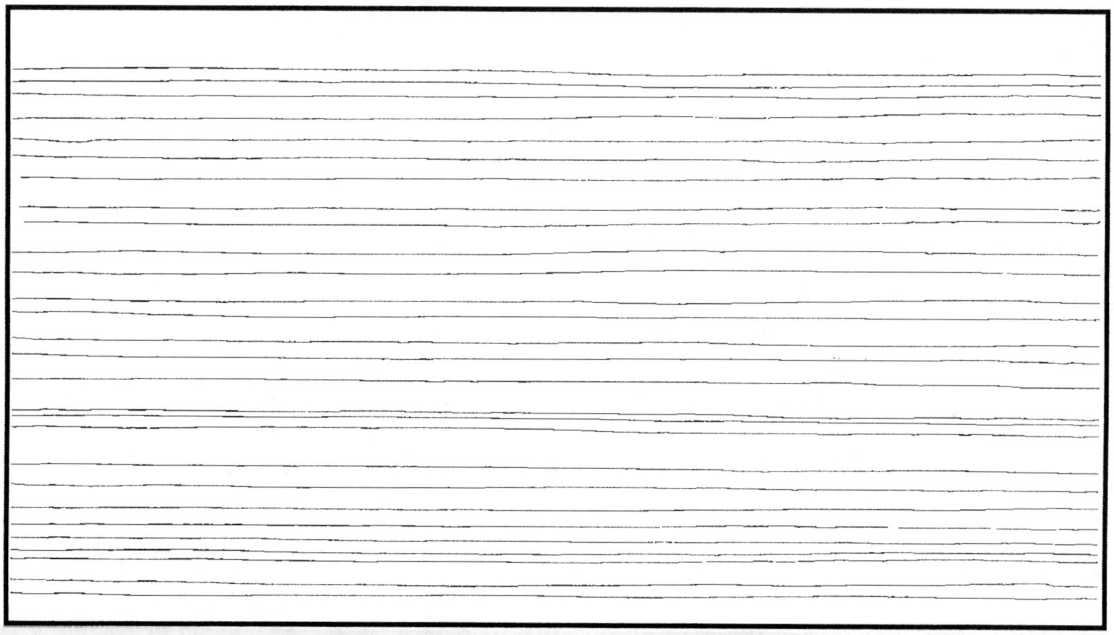

Horizontal Lines

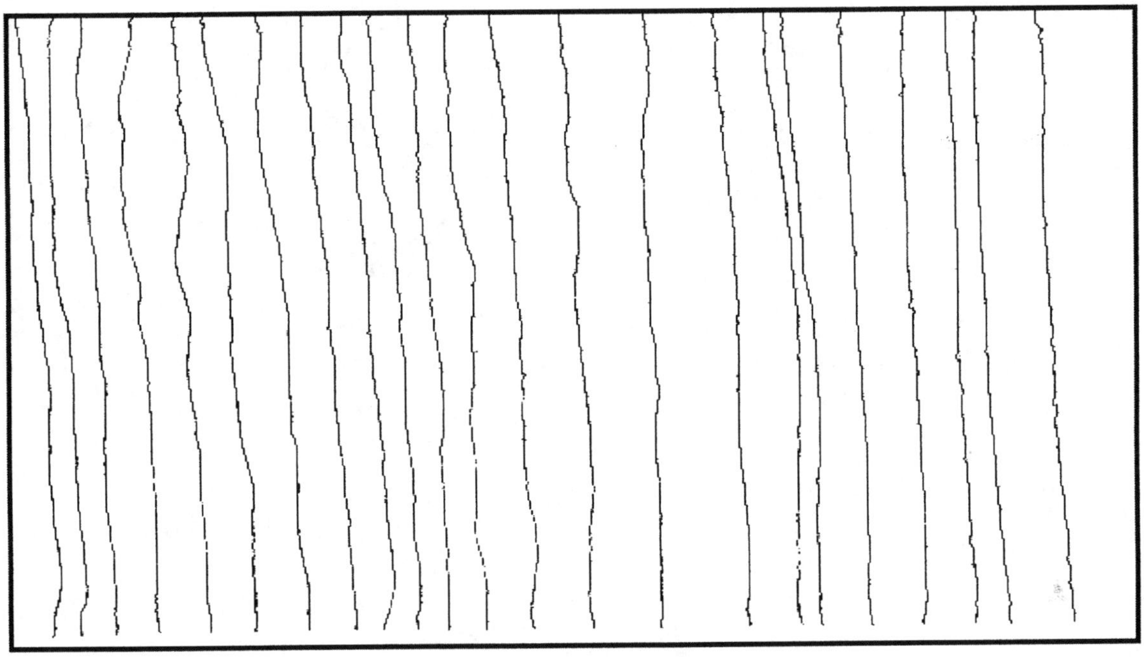

Vertical Line

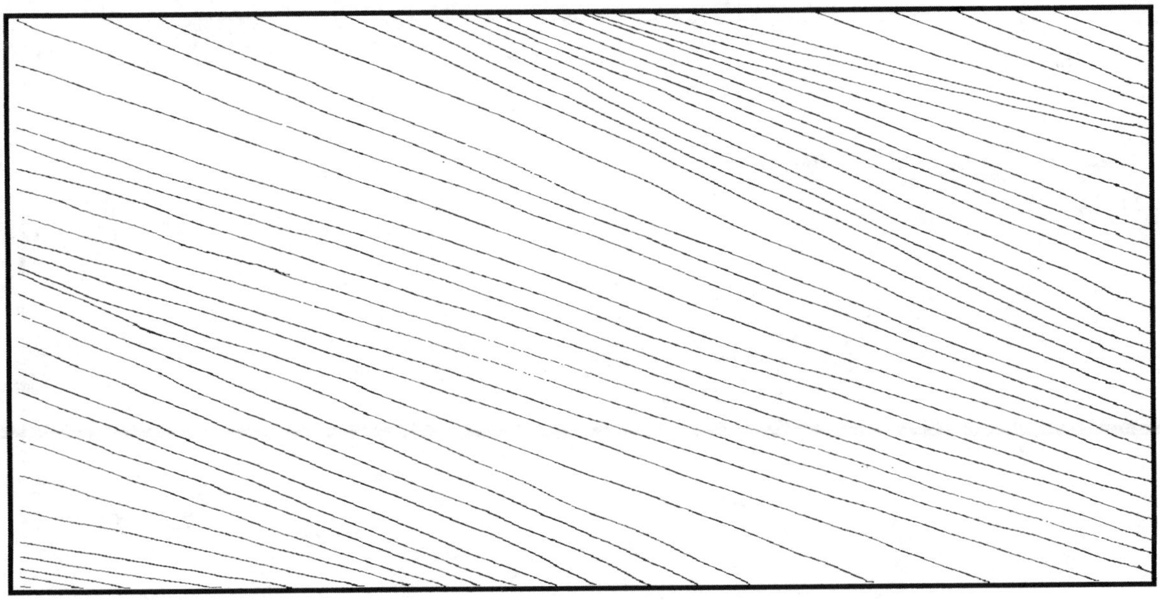

Diagonal Lines

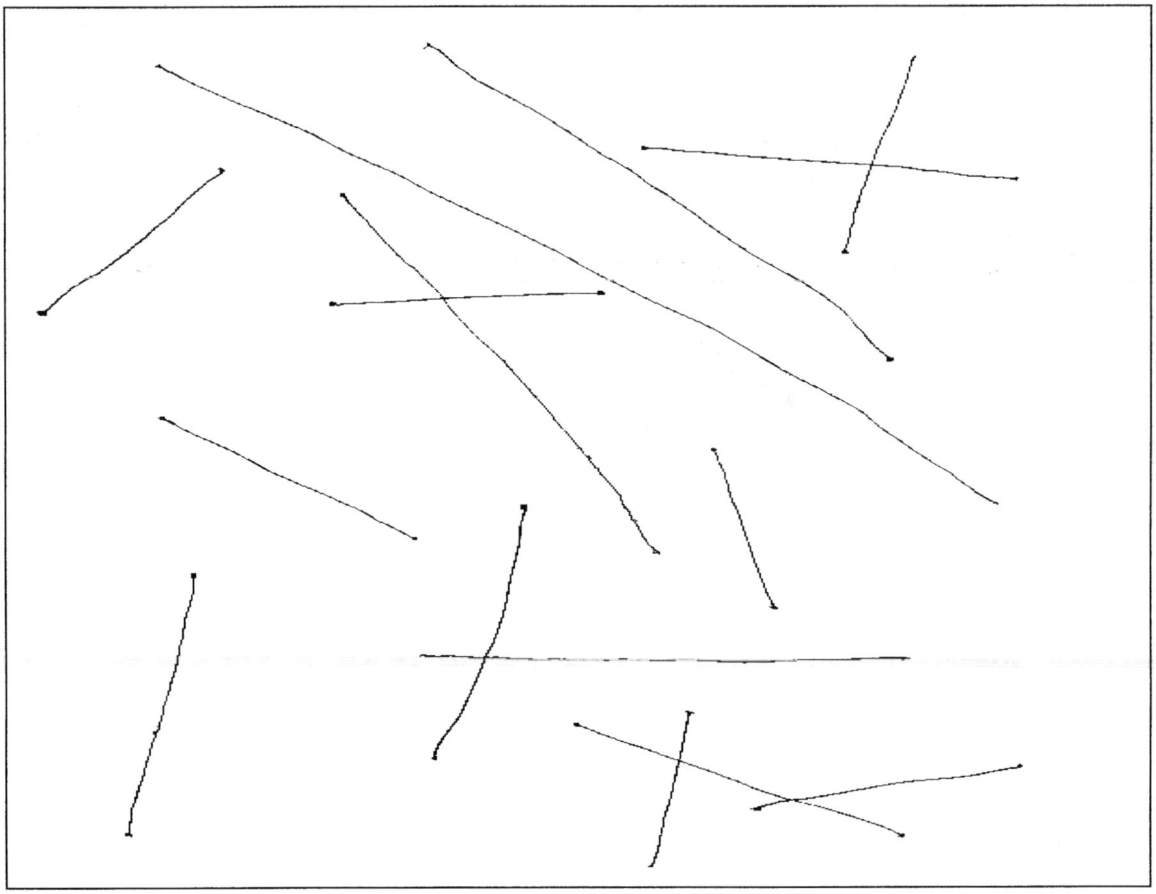

Lines Drawn “Dot-to-Dot”

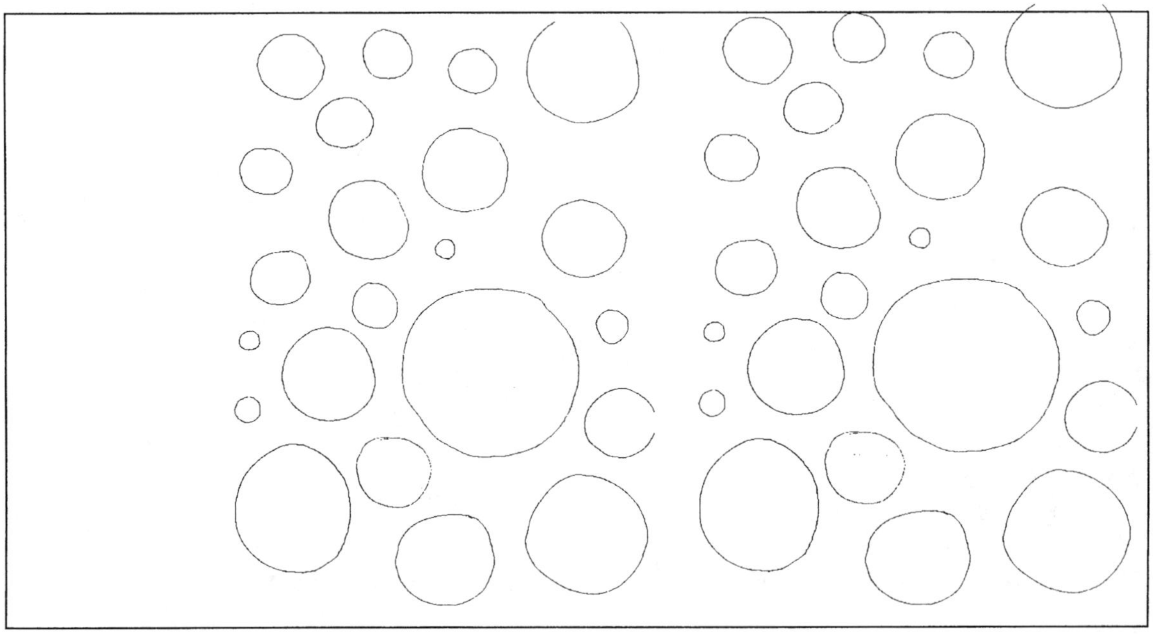

Circles

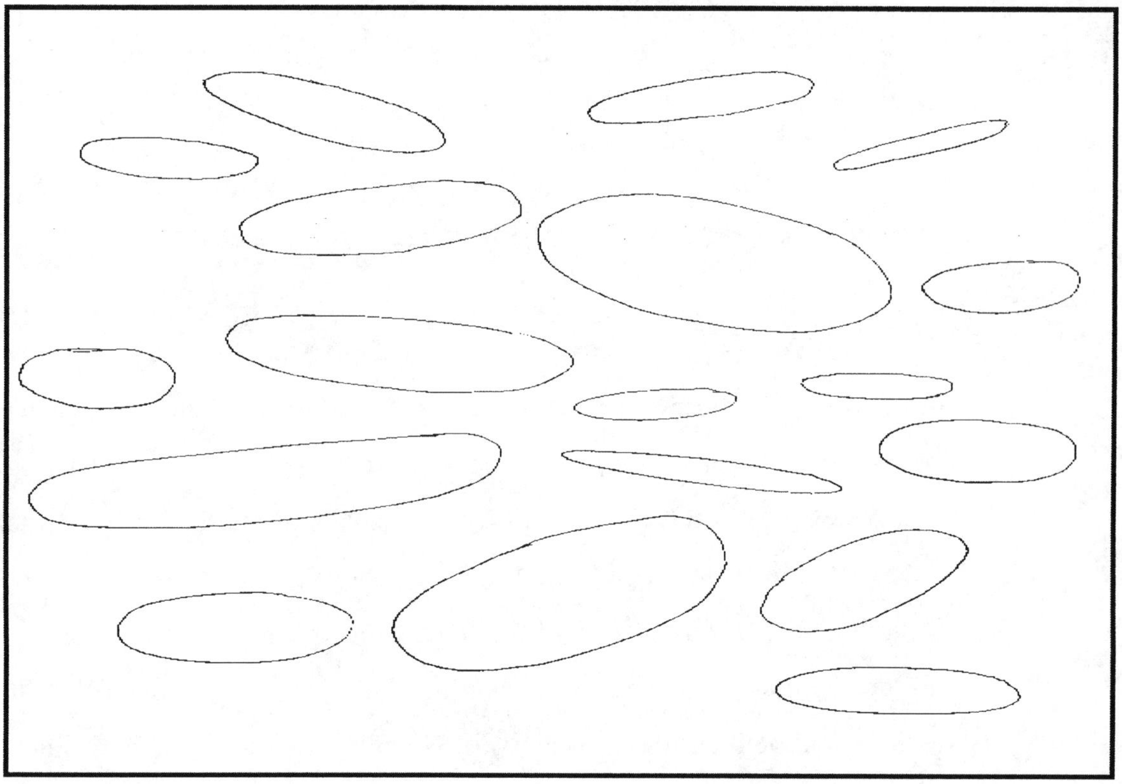

Page of Ellipses

Chapter 4 - Contours

The exercises outlined in the next few chapters build the skill that you need to "see" in the way an artist sees, and to attain the visual processing skills needed to make competent freehand sketches. They are designed to train the eye and the mind to focus on visual information without the interference of what the left brain knows about the object being studied. It is important in order to gain the most from the series of exercises to follow them carefully and to do them in order. The degree of skill built will depend both on how closely you follow the directions and how much time is spent on the exercises.

Blind Contour

It is important that you focus on the process of seeing rather than producing an end product or "drawing". To that end, we will refer to these exercises as studies. Remember, and this point cannot be emphasized too much; it is the process of seeing that is being worked on now, not the ability to produce a drawing. To accomplish this, it is also important to refrain from evaluating either the process or the result, so one of the rules is "do not erase". Erasing signifies the desire to get it "right"; which is not the point to these early studies. Such evaluation can interfere with the process of "seeing". If necessary, remove the eraser from the pencil you are using or use a pencil without an eraser. It is also important to keep from thinking about the object that you are studying and to focus on the visual information that it contains. Any time the name of the object or some fact concerning the object enters your mind, pause, and let it pass. Thinking about what you know about the object is left-mode processing and will prevent right-mode processing from occurring.

Select an object to study. Natural objects work best. They generally have more contours and less is known about them. One of the more common objects for this first attempt is the hand, so let us start with that. Read all of the instructions before you begin, then follow the instructions carefully.

1. Take a clean white sheet of paper and tape it to a table top. That way, the paper will not slide around as you work. Make sure that there is a pad under it so that the pencil point does not tear the paper. It is also important to keep distractions to a minimum.

2. Lay the hand you are going to study in your lap with the palm side toward you. Turn so that you are facing away from the table and the piece of paper. We will turn the palm up for this study since there is less symbolism associated with the palm (of course you can include the fingers). Let your fingers relax so that they curl toward you. This view also produces more contours.

3. Now turn your hand and head away from the paper so that you cannot see what the pencil is doing. This is why the process is called blind contour. Look at your hand; that is where the information is that you are going to record. To make it more interesting, take the pencil in your non-accustomed hand. It really makes no difference what hand you use, because it is the eye and the mind that are doing the really important work.

4. Do not be concerned with what the pencil is doing or how this is going to turn out. That is not important, only the process of seeing, of processing the contours visually is important. If the pencil should come off of the paper, simply move it back and continue. You are going to record this experience in one continuous process, so keep the pencil in

constant contact with the paper. Unless it comes off the edge of the paper while recording, this will be one continuous line.

5. You will not look at the paper at any time during this process. That is not where the information is, it is in your hand. Look at your hand. Notice the lines formed by contours and edges. Take some time to study them. Notice how they curve, how they relate to one another. Chances are that you have never studied your hand in this manner before, so try to learn something about your hand that you did not realize. Relax and get comfortable.

6. It does not matter where you start. Wherever your eye happens to land is fine. You are going to work as though your pencil point and your eye are attached mechanically. Where your eye goes as it traces the contour, the pencil point will follow, at the same rate, and in the same direction. Move slowly; study each minute variation of each contour and record it faithfully

7. It is important not to evaluate. If a contour does not go where you think it ought, ignore it and continue. Under no circumstances should you erase a line. That will take the focus off of visual processing and also means that you are looking at the paper. The point is to process in a purely right-mode. No words, symbols or evaluation. All of those are left-mode and will interfere with the right mode. It is not important that you record all of the contours. Continue in this manner for about 2 or 3 minutes the first time. Do not worry about time. Use a bell type egg timer if you have only a limited amount of time to spend at one sitting. Concern with time is also a left-mode process.

8. When you have finished, look at the paper. You may have been dying to do this since beginning the process, but that urge will pass with more practice. What does it look like? The question is a natural one, but an unimportant one. It looks like a blind contour study. Some of the

essence of the hand may be there, but if it looks too much like a hand, maybe you were a bit too concerned with the results.

Again, it is important that you follow this process just as described. Do not give in and look at the paper, no matter how great the urge. The information that you need to draw is in the object, not on the paper. This is the first step in learning the discipline to focus on spatial information and process it visually. It is an all important step in learning both to visualize and to draw. How much you progress depends on how closely you follow instructions, but as mentioned before, it also depends on the time you allow. Repeat this study for a total of two hours or more before going on to the next type of study. It is important to spread the two hours over several drawing sessions (about 30 minutes). Begin by repeating the hand exercise, but this time with your accustomed hand. You will notice that it really does not make any difference, since it is seeing that really matters. Natural objects work best, such as pine cones, plants, bark, shells, grass, or anything that is interesting, full of contours, and available. Manufactured objects do not work as well because the contours are usually straight and boring. Occasionally, an object like an egg beater, whisk, or complex tool will also work as long as there are lots of contours.

It is important that the time spent in completing any visual study be undisturbed. Make sure that you are alone, the phone is off the hook or disconnected, and that all other distractions are at least minimized. It is also important that you are relaxed. Do not attempt these studies if you are anxious or under a lot of stress. It will not work if you have just had an argument with some one or if you just got home after driving through some horrendous traffic. Relax first. Meditation is one of the best ways to relax. Sit somewhere quiet and let all of the cares melt away. You will find that these studies will eventually aid this process. Many artists paint or draw just for that purpose. Winston Churchill credits his painting with getting him through WW II without having a mental breakdown. Give it a chance.

Pure Contour - Piece of Paper

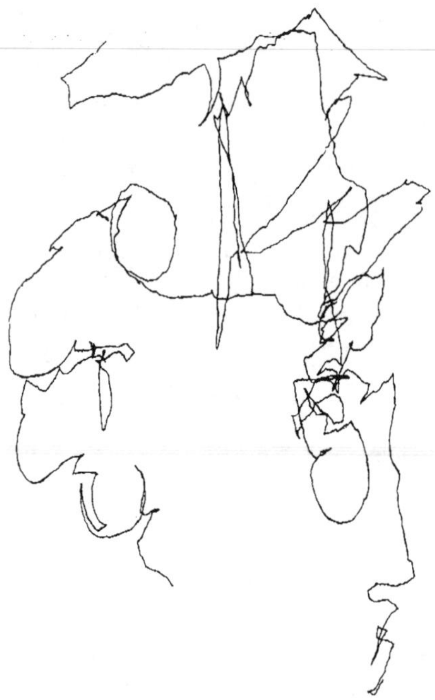

Pure Contour – Hand

Modified Contour

The main benefit of blind contour study is in training the mind and eye, and as one develops competence in seeing, it is used primarily as a warm-up exercise. Many artists use it as a warm-up prior to beginning a drawing session in order to make the right-mode shift to visual processing. A more useful process, and the next step in our progression toward seeing, is similar to blind contour with the exception that we will occasionally glance at the paper, not to see what we are drawing, but to allow more visual information in. Again, begin with the hand. Read all of the instructions before you commence drawing.

1. Start the same way you would when doing a blind contour with the hand held loosely in the lap, the palm up, and a clean sheet of paper taped to the table. Face away from the table and paper, but at an angle such that you may comfortably look around at your work.

2. Before you begin, study the hand; continue to try and discover something that you had not noticed before. Concentrate on the contours. Notice the visual patterns that they make; the shapes and angles; the lengths, direction and proportions.

3. Again, you may begin at any place you like. As in blind contours, make a continuous line, keeping the pencil point in contact with the paper the entire time.

4. As you move your eye over each contour, studying each tiny variation, your pencil should move as though attached to the eye by some invisible arm. Move slowly, study carefully, and record the contour. Remember that concern with time is an enemy of right mode processing.

5. Here is where the modifications come in. When you reach a point where the contour makes a sharp deviation, say at the knuckle, pause and take a quick look. Note the proportion of the contour just recorded to the next contour; the angle, the shape it makes. Then continue recording. While recording, the eye must be on the contour, not the paper.

6. Since you are looking at the paper occasionally, the urge to get it right may be even stronger than with blind contours. Resist this urge, since it involves evaluation, and that would be letting the left-mode back in. If a line is in the "wrong" place, do not attempt to erase it, just restate it with another line. The whole purpose of this exercise is to learn to process in the right-mode, blocking the left out.

7. A name associated with the object you are studying may pop into your head. Pause and wait for it to leave. Names evoke your stored memory and you may end up recording what you know about the object rather than what you see in the object.

8. When you have recorded all of the contours, stop and look at your work. You will notice that there is a much stronger resemblance to the object than in blind contour. It is all right to notice this, but do not become concerned with it. You are not trying to make drawings, but to learn to see with the aid of modified contour studies.

Again, following instructions is important, but the only way to learn to draw is by drawing. You will use this technique for many of the other studies you will be doing, but it is important to work with it now. Work on studies for at least 2 to 4 hours in roughly 30 minute sessions before going on to the next chapter. Almost anything will do for these studies. Natural things have more contours and are more interesting most of the time, however, an old pair of shoes, an old coat, or a book bag can be just as intriguing. If you are so inclined, tools also can work well. The point is, draw!

Modified Contour - Hand with Flower

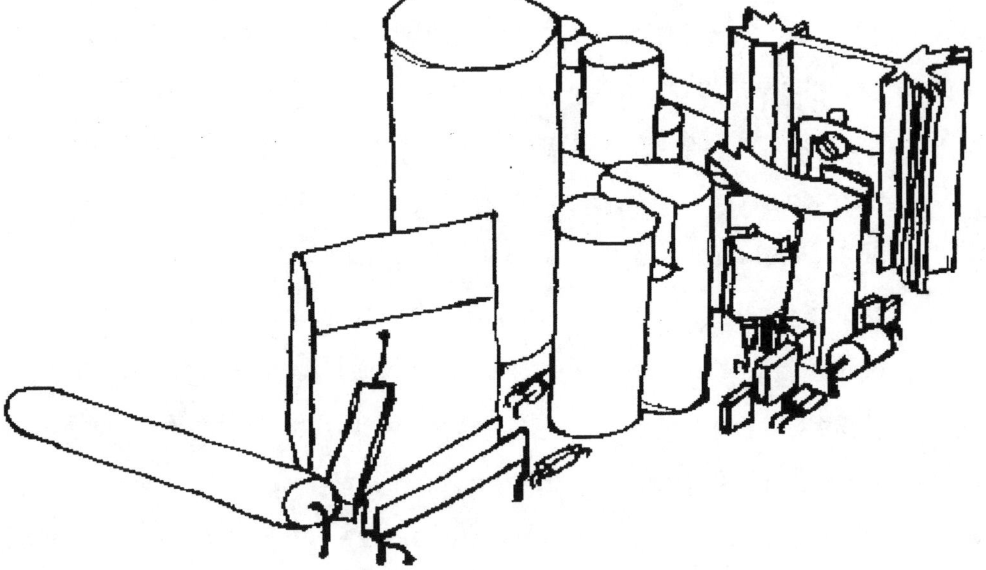

Modified Contour - Computer Circuit Board

Gesture

All of this no peeking and slow contour processing can tend to make one a bit up-tight. It is now time to loosen up a bit with some gesture studies. The purpose of gesture studies is to capture the essence of shape and form as quickly as possible. In a sense it is the opposite of contour drawing, which is to capture visual detail. The ultimate purpose of both techniques is to develop the ability to process information visually; one for detail and one for basic form. Gesture is also valuable for capturing fleeting form, such as that involving movement, or to capture the form of moving subjects

The process for gesture studies involves capturing the form quickly, so the focus needs to be holistic, rather than detailed. Detail is not important, only form or mass. To do this, a full and connected arm-shoulder motion is preferred. As in contour studies, the focus is on the subject and not the paper. The entire gesture study should be completed while focusing solely on the object. The studies are of short duration, 10 to 20 seconds maximum. When studying a bicycle that is passing by, for example, the subject is seldom present for any longer period of time.

Gesture studies are done as follows. First, as always, read all of the instructions before beginning the study. The first several studies are best done while standing, preferably at a tall table, so that the arm and shoulder can be involved. Pick an object for the study. Start small. The object can be either stationary or moving, although it is probably best to begin with stationary ones. It is also better if the object is rather large, say a tree or a large piece of furniture.

1. Study the object momentarily before you begin, ignoring detail but seeing the overall shape.

2. Use large circular motions to duplicate the shape/form that you are observing. Do not look at your paper. This is one continuous motion, so keep the pencil (or other implement) in continuous contact with the paper.

3. Capture the form quickly. Set a timer and work for 15 seconds or less. There are no rules for what to do first, simply draw what you see, quickly focusing on form rather than detail.

4. Use the same technique for moving objects such as a passing car, bicycle, or person walking.

Gesture – Sitting Under a Tree

Gesture - Person Walking, Bike Rider

These studies can be done quickly, so many studies can be done during any drawing period. Practice on both organic and inorganic forms. As before, practice is the most important aspect of learning to draw. Gesture studies are particularly adaptable to out-of-doors. You may want to sit on a bench in a park or along a street and watch the people pass. Capture their fleeting forms, the movement of their limbs. Bicycles and other slower vehicles also work well for gesture studies, but then, so do buildings and trees. The point is to loosen up using form and mass as the visual focus rather than contour.

Gesture - Building With Surroundings

Chapter 5 – More Ways of Seeing

Negative Space

Visually, any object consists both of the contours that make up shape, form, and detail, and the contours that make up the space within and surrounding the object. As for example, a wicker chair is made up of the contours defining the object and also the space seen between and within the parts making up the chair. They are both separate and at the same time one, like a cup and the space within the cup are one. Although they cannot be separated physically, they can be separated visually. One can "see" the object by seeing its negative space as well as one can see an object by studying its contours and form. Seeing space can be helpful in seeing the form, particularly when studying objects which hold strongly embedded symbolism, i.e. those objects which are most familiar to us. If one is having trouble with the contours defining the form, one can switch focus to the "space" around and within the form. Space seldom holds strong memory images which might interfere with the "seeing" of the contours. In addition, developing the ability to visually process negative space shapes adds one more skill to our visual processing repertoire.

Negative space studies are most successful if they are done on objects with lots of interior space such as a chair, particularly one with slats and leg braces, or plants, a random coil of rope or string, tables, or tables and chairs together, to name a few. Wicker or oriental chairs make ideal subjects, as does anything that has lots of interior space within its outer boundaries.

Viewfinder

One of the problems inherent in negative space studies, is the lack of a boundary for the space surrounding the outside of the object. Theoretically, this space is infinite although it is bounded somewhat by our immediate surroundings and our field of view. Even those boundaries make dealing with the surrounding space problematic. What is needed, at least in the beginning, is something for bounding the surrounding space. That is what a viewfinder is for. This consists of a piece of stiff paper with a rectangular hole cut in the middle. You may hold this device between you and the object such that the object is surrounded by the opening in the same manner a matte might frame a picture. Before you begin negative space studies, it is suggested that you make a simple viewfinder. The material for this device should be a heavy piece of paper, something that will remain upright while holding it at the bottom edge; a piece of card stock or matting material. A proportional rectangle needs to be cut in the center of the material approximately 1.5" along the long side. This can be accomplished in the following manner.

1. Use a piece of heavy card stock which will stay upright when held at the bottom edge.

2. Mark a pair of diagonal lines across the paper.

3. Using a straight edge, make a line intersecting two of the diagonal lines, parallel to the long side of the paper and 1.5 inches in length.

4. Where this line intersects the diagonal lines, draw a second line at the upper intersection and parallel to the top of the page, and a third line starting at the lower intersection with the diagonal and parallel to the bottom of the page.

5. Next, connect the unconnected lines with another line parallel to the side of the page forming a rectangle.

Using the diagonals will make the inner rectangle proportional to the paper. Lastly, cut out the rectangle forming a window or view port. (See illustration on page 35)

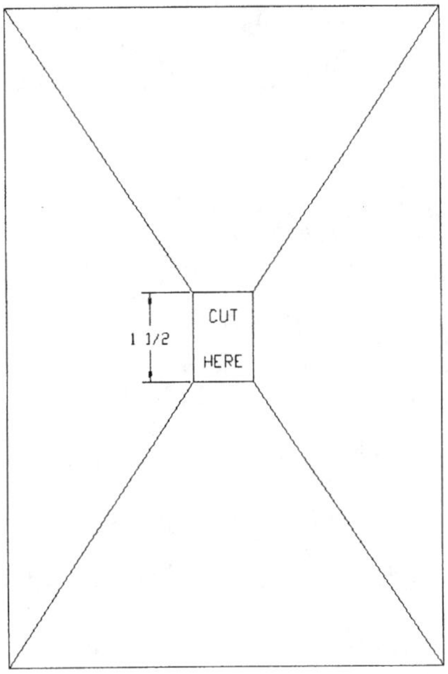

Viewfinder

Negative Space Studies

You are now ready to commence making your first negative space study. You will use a process similar to modified contour, only this time you will be focusing on the boundaries of the space within the object. Choose an object with an abundance of space, such as a wicker chair, or at least a chair with lots of rails between the legs and at the back. Other objects such as plants, a piece of string wound around randomly on a flat surface, or an eggbeater, etc. will also do. It is best to start with a large object. Place the object on a table (even if it is a large object) so that the center of the object will be at approximately eye level. This allows you to continue the study for a long period of time without tiring. Use the following procedure to complete the study. Read through all of the instructions before you begin.

1. Hold the viewfinder in front of you so that you can see the object of the study. For the sake of clarity, let us assume that the first study is of

a chair. Move the viewfinder back and forth until the sides of the viewfinder visually make contact with the chair on at least three of the edges. Before you begin, study the chair carefully, picking out the space in between the various parts of the chair. Study these spaces until you can clearly see them as individual shapes. Study each space in turn until you see each as a distinct shape. It is these shapes that will constitute the study. Notice that the shapes are independent of each other. They do not touch; each shape is a necessary part of the whole.

2. Pick a shape near the outer edge of the chair. Using the modified contour technique, carefully study and record this shape. Use the edge of the viewfinder to help see the visual and spatial characteristics of each line in the shape, the angle it makes with the edge of the viewfinder, its proportional length, and its curvature or straightness. Duplicate these spatial conditions when you record the line.

3. As in modified contour, work slowly and deliberately, capturing the unique qualities of the edges of each space shape. Remember, you are not in a hurry and you are not producing a product. The purpose here is to develop your ability to see visual form.

4. Next pick an adjacent shape. Use the first shape to add more visual information for locating and recording the second. Notice the relationship between the two and the distance between them. The distance between space shapes is the positive part of the object.

5. Continue this process, one shape at a time, like putting together a puzzle. Study and record each shape carefully, capturing all of the visual and spatial relationships that make up the whole.

6. After you are finished, take a look at the finished study. The set of shapes are not the chair; you have recorded no contours, only space; yet the negative space captures the essence of the chair. The chair and its space are one.

If you have been successful in following these instructions for negative space studies, the contours which make up adjacent spaces should not connect. If they do, this represents a shift from focusing on negative space to focusing on the contours that make up form.

This process can be used whenever you are experiencing difficulty with seeing the contours of an object, particularly one with which you are intimately familiar, such as the features of the human face. If you are having difficulty seeing a contour, i.e. what you know is interfering, switch to the space surrounding the contour. There are no memory symbols for negative space shapes. Complete several studies using a variety of objects. The only requirement is that there be plenty of negative space.

Negative Space - Chair

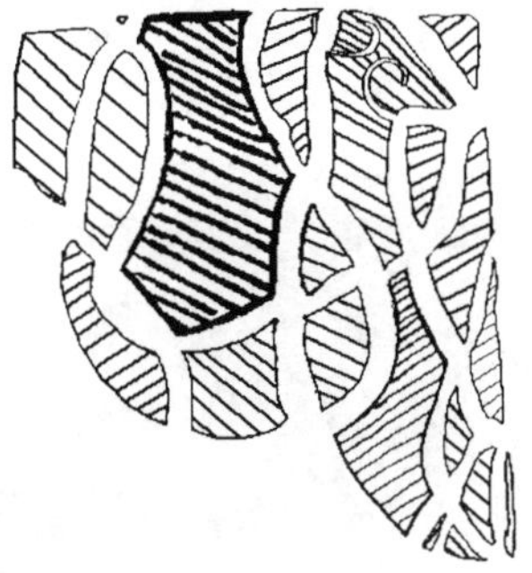

Negative Space - Rope

Negative Space - Stack of Chairs

Chapter 6 - Perspective

One of the characteristics of the studies that you have done so far is that they are all rather flat looking; there is no appearance of depth. Also, you may have noticed that it is sometimes difficult to get the proportions just right. We will now work on a technique which will help you solve both of these problems, that of "sighting". First, it is important to understand how the mind perceives three dimensional space in a two dimensional medium.

For many years, scientists and doctors attributed our ability to perceive 3-D space solely to binocular vision. More recently, they have discovered that the brain uses a number of clues to perceive space. We will not go into all of them at this moment, but it is important to mention some of those which are visual in nature. The most important of these are size diminution, hierarchical order of objects in space, overlapping of objects, loss of detail, and loss of color saturation. We will take each one in turn, beginning with size diminution.

The word diminution means making smaller or appearing less. The best example of this phenomena is the reference made to people as appearing like ants when viewed in the street from the top of a high rise building. Objects far away appear smaller than objects close by. Size diminution leads to another visual effect of 3-D space, that of converging lines. The sides of a road or the rails of a railroad track seem to merge as they are viewed from near to far. The same is true with any set of parallel lines or edges (contours) of objects when viewed from an angle resulting in the object starting close to the observer, then getting farther away; the sides of a building, a street, a sidewalk, or any large object. This effect is due to the size diminution of the space between the parallel objects.

Actually, the phenomena exists when viewing any object, no matter how small, but is more easily observed in large objects. When this phenomena is incorporated into a drawing intentionally, it is referred to as perspective. A related 3-D visual clue is loss of detail. As objects appear to get smaller, the small details become more difficult to distinguish or seem to disappear altogether.

Another important visual clue to 3-D space is the hierarchical order of objects. Those objects close to the observer appear lower in the visual field, and objects further away appear higher. The best example of this visual phenomena is to observe a large room full of many objects such as tables and equipment in a large drawing room or computer lab. If you stand in one corner and look across the room, those objects near to you will appear lower in your visual field than the objects at the far end of the room. In this same room, it is possible to also observe the phenomena of overlapping objects. Those objects nearest the observer visually obstruct the view of the objects behind them.

The last clue mentioned here has to do with color and will become clearer after reading the chapter on color. One of the dimensions of color is color saturation. As colors blend visually, it becomes more difficult to distinguish hue, or just what color objects are. This phenomena is caused by particles in the air which tend to combine the various colors. If we look at objects several miles from us, the color of different objects is indistinguishable. Thus the reference to “purple mountains”. The “purple” referred to is actually gray and is lighter or darker depending on the amount of moisture or pollution in the air and the lighting conditions.

Visual sighting can also aid in overcoming the phenomenon of size constancy. This happens when our mind tells us that objects far away are really the same size as objects close by. It is based on information stored in our brain about the nature and characteristics of those objects as will be shown later in this chapter.

Sighting

We will first learn a technique which aids the mind in dealing realistically with what the eye sees. This is the technique of sighting, frequently used by artists to accurately incorporate size diminution (visual perspective) and proportion into their work. A long office pencil works best for this purpose. Three rules apply; the pencil must always be

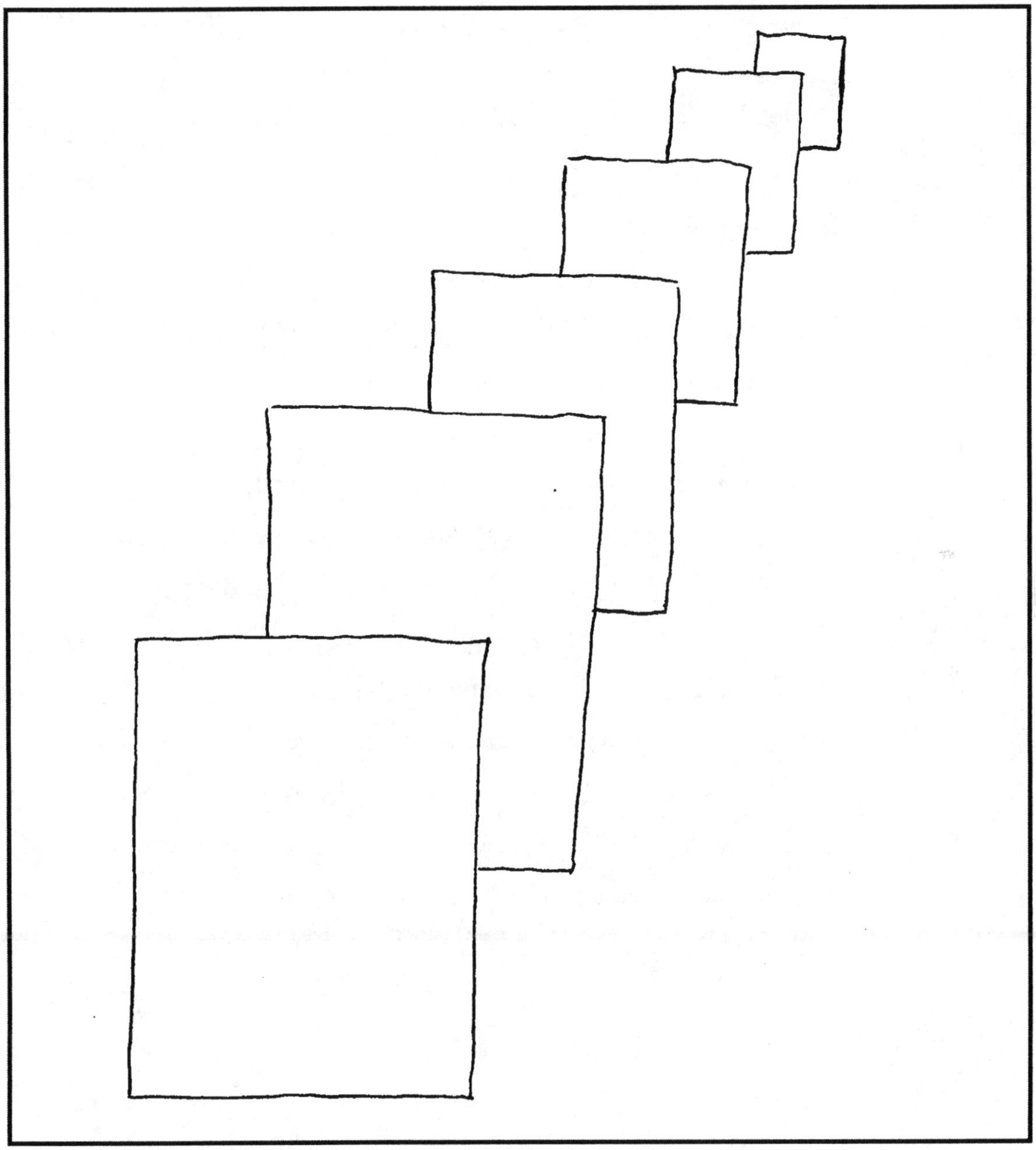

Demonstration of Depth Clues

held at arms length, it must be held either vertically or horizontally, and it must be held so that it is perpendicular to your arm. The visual clues we are seeking are 3-dimensional, however, we are recording them on a 2-dimensional medium. Following these three simple rules is vital in making this transfer from 3-D perception to 2-D medium an accurate one. A quick exercise to demonstrate the potential conflict between what the eye sees and what the brain or memory/symbol system tells us is to compare the size of human heads; those near to those at some farther distance. When you are in the classroom, or any room with several people, look around and see if there is anyone with an unusually large or small head. Most likely there is none or at least this is what your brain and memory about the size of peoples' heads tells you. As mentioned earlier, the psychological phenomena of size constancy tells us that similar objects are of similar size whether near or far from the observer. This is not what the eye tells us. To illustrate this point, first of all focus on the person nearest to you. Position your pencil at arm's length between you and that person. Position the pencil such that the tip of the eraser is even with the top of the person's head. Now slide your thumb down the pencil such that the tip of your thumb is even with this person's chin. Holding your thumb in that position, move the pencil to a position between you and the person the farthest away in the same room. Move the pencil so that the tip of the eraser is again even with the top of this person's head. Note the relative or proportional difference between the thumb and this person's chin. The head further away may appear to be 3-4 times smaller than that of the closest person. We know instinctively that these heads are not that different in size and this knowledge can create a conflict between seeing and knowing which prevents the making of an accurate study or the accurate recording of visual information. This is where the sighting technique comes in.

For this first sighting study, use the chair or object that you used in the first negative space study. Again, set it on a table, this time a low table is preferred. Sit in a comfortable position with your paper in front of you. We will begin by establishing the overall horizontal and vertical dimensions of the object. Hold the pencil at arm's length between you and the object of the study, beginning with the pencil in a vertical position. If you are not sure that it is vertical, sight a corner of the room, and align the pencil with the corner. Position the pencil such that the tip of the eraser is even with the top of the chair. Now position your thumb along the pencil such that the tip of your thumb is even with the end of the leg lowest in your visual field. Mark this distance on your paper. This can be done by transferring it directly from the pencil and thumb to the paper. If direct transfer is used, it is important to select a long enough pencil and to position yourself at a distance from the object (chair) such that the study fits comfortably on the page, filling most of the page. Also, take note that the legs do not all appear at the same level or to be the same length. The visual phenomena of size diminution and hierarchical order are at work here.

Next turn the pencil to a horizontal position, keeping it perpendicular to your arm and at arm's length. Position the pencil such that the eraser tip is even with one edge of the object. Slide your thumb along the pencil until the tip of the thumb is even with the other side of the chair. Mark this distance on the paper. You will use this technique to establish the proportional relationships for all of the respective parts of the object.

Take a look at the contours of the chair or object. Not all of them are vertical or horizontal respective to your visual field. To determine the visual angles accurately, hold the pencil either vertically or horizontally, whichever is most appropriate. Move it to a position relative to a contour defining an edge of the chair. Note the angle that the contour makes relative to the position of the pencil. This too can be transferred to the paper.

You are now ready to begin a modified contour study of the chair. When you are recording a contour, there should be no difference with the earlier contour studies you have done. Keep your eye on the contour and record it as if the eye and pencil were one. The difference is that you will use the pencil and the sighting technique to establish the spatial qualities of distance, angle and proportion. It can also be used to estimate curvature, and the relative position and size of arcs and circles. Begin recording at either the top or bottom and work through the object, letting the contours your have recorded aid in the positioning of new contours. Work slowly and deliberately. Remember that concern with time is a left-mode process and can interfere with the right-mode process of seeing and recording visual/spatial information. Continue this sighting and recording process until you have recorded all of the contours in the chair.

Sighting Study Of Chair

Stack of boxes

For this study, you will need several large, several medium sized and several small boxes. This is one of the more challenging studies for sighting in that boxes carry a great deal of learned symbolism. We know what boxes should look like. To somewhat confound the brain and make the study more challenging, create an interesting stack of these boxes on a table. The table can be a low one, and you may be able to see the tops of most of the boxes, but be sure to make the stack high enough so that the bottoms of some of the boxes are also visible. Turn the boxes at various angles so that none of the sides nearest to you are parallel to any other, and position the boxes such that some of the corners stick out, overlapping the front edge of the box above or below it. You are now ready to begin. Be careful. We know a great deal about boxes and what we know can interfere with what the eye sees. The trick here is to focus on what you see, the contours formed by the corners of the boxes. Use the sighting technique to aid you in establishing the various visual relationships such as proportion, angle, distance, shape, etc. Use the modified contour technique to record this information.

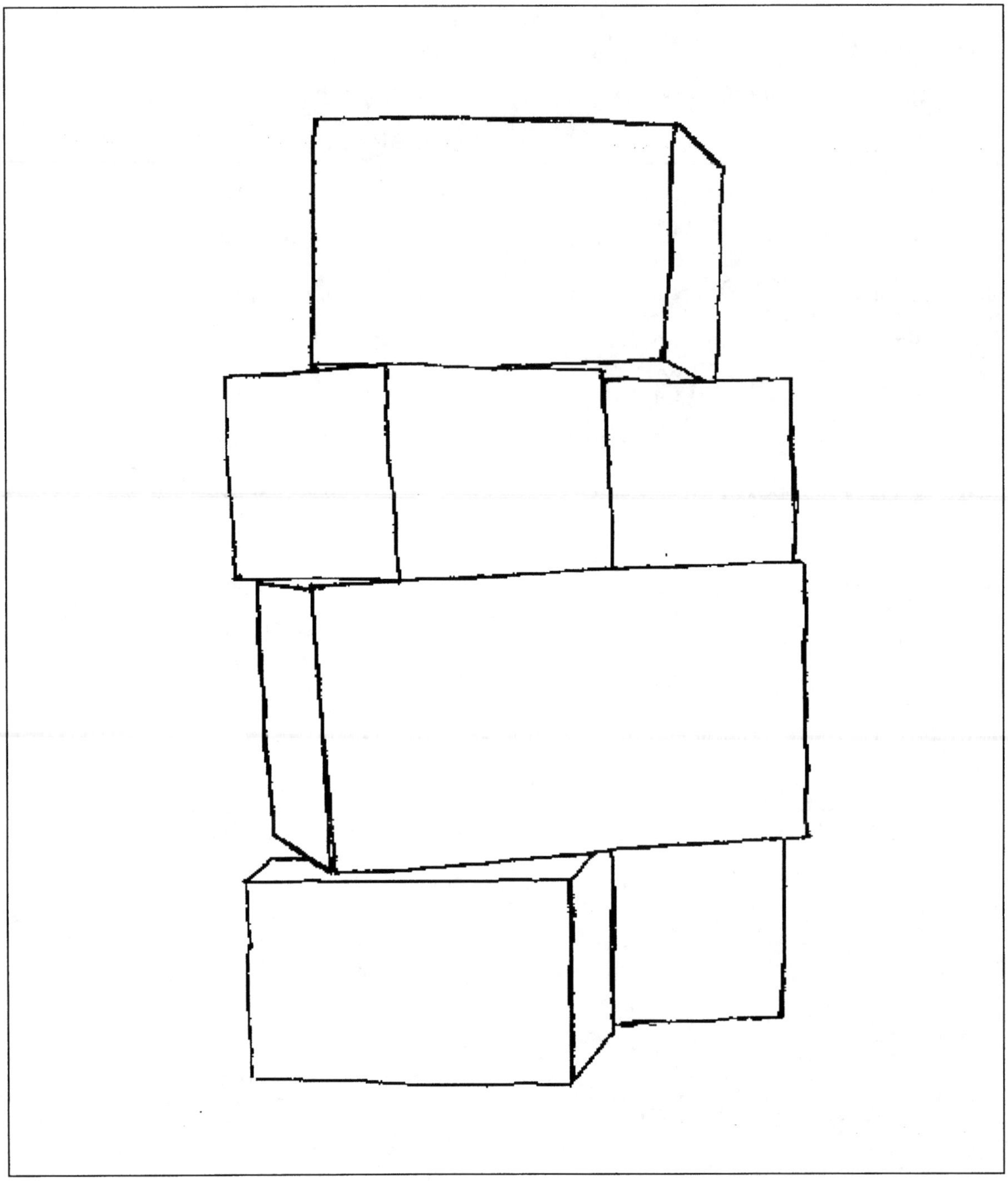

Stack of Boxes

Room Corner

The ultimate challenge for a study using the sighting technique is a room corner, preferably a kitchen looking into the corner with the most cabinets. Cabinets and counters offer a large number of contours and edges to process and some challenges between seeing and knowing.

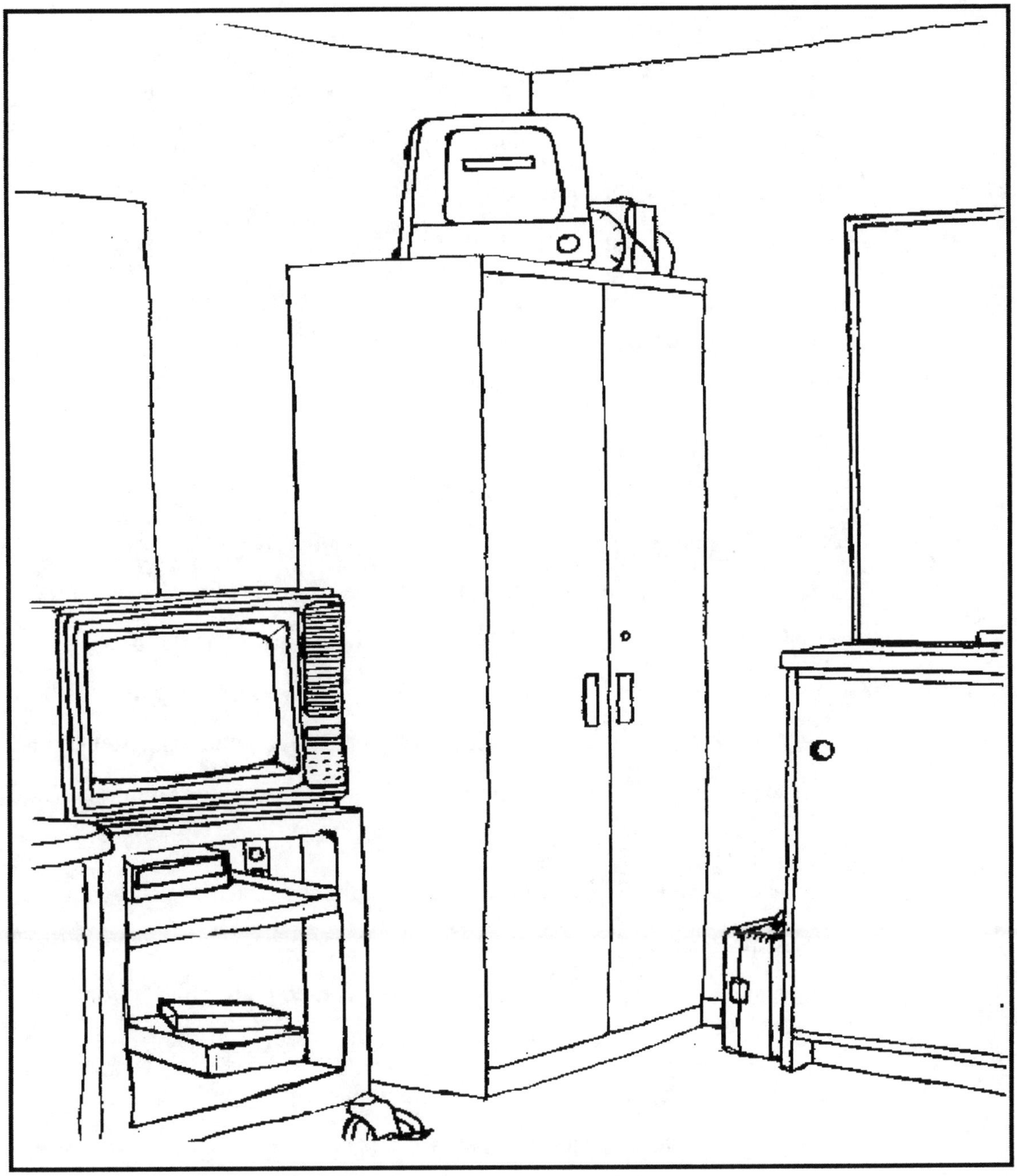

Busy Room Corner

One, Two, And Three Point Perspective, A Quick Look At The Formal System

The formal system of perspective was developed by mathematicians to overcome the problems inherent in trying to see the aspects of perspective, namely size diminution, with the eye. Artists, beginning with Albrecth Durrer in the 16th century, worked at developing various techniques which basically all involved some sort of sighting device which aided in incorporating line convergence and other aspects of depth. After the mathematicians created a rigid formal process for incorporating perspective, artists abandoned the process and the age of expressionism was born. The biggest problem with formal perspective is that the vanishing points are usually placed artificially close together. This is necessary in order to get them on a reasonably sized piece of paper.

There are three ways to incorporate perspective in the formal system; one-point, two-point, and three-point. The "points" in this system theoretically occur where the principle of size diminution causes the object to disappear, or "vanish". If these points are to be located on a piece of drawing paper or table top or can be drawn with averaged sized straight edges, then the vanishing points will be artificially close to the object, causing an exaggerated line convergence effect. However, the system does give objects the appearance of great size which can be useful for certain visual effects, particularly in producing architectural renderings.

The two-point perspective process is the most frequently used, mostly because it can incorporate a wide range of visual effects with relative ease. One-point perspective is used for viewing an object from its interior, such as a long hallway. Three-point is useful when trying to create the effect of viewing a very large and particularly tall object such as a sky-scraper building or tower. We will review each of the three types of perspective in a very brief and simple manner, using only visual techniques and a few basic concepts, rather than the complete formal structure of the system.

The cube and the right rectangular prism are the most basic of geometric forms incorporated into our man-made world, and is the most useful for demonstrating the different types of perspective drawings. This form is also useful in creating "boxes" in which other solid forms may be drawn. The first exercise will consist of a series of cubes in one-point, two-point, and three-point perspective.

Two Point Perspective Begin by turning your drawing paper in the "landscape" orientation, and use the following procedure:

1. Draw a horizontal line across the page just slightly above the center. This line represents the horizon line. Any object totally above it will appear to be floating and any object totally below it will appear to be below the viewer.

2. Next, place a point on this line at either end next to the edge of the paper.

3. Begin the cube with a vertical line drawn anywhere between the two vanishing points. It does not really matter if it is above, below, or on the horizon line.

4. Next, draw a very light weight line from each end of this vertical line to each vanishing point (four lines in all).

5. Visually judge approximately where a second vertical line should be placed parallel to the first such that the enclosed area approximates a square. Do the same for each side of the first vertical and draw both of these lines.

6. Draw additional light weight lines from the ends of each of these new lines back to both vanishing points on either side. Darken in the edges of the cube, judging which lines will be visible.

7. Create a series of cubes in different places on the paper, some on, above, and below the horizon line. Begin each new cube with a vertical

line which represents the corner nearest to you, and then follow the instructions as before.

Two Point Perspective

Three Point Perspective Three point perspective works much better if you use a larger piece of paper, at least 18" X 24" (C-size). Place the horizon line in the lower half of the page with two of the vanishing points at either end. Place a third vanishing point near the center and as far up the page as possible, or even on a piece of tape off the page. Draw a long, rectangular prism, imagining that it is the outline of a very tall building and use the following procedure:

1. Begin the drawing with a vertical line drawn anywhere between the two vanishing points. Begin below the horizon line and continue to near the top of the page.

2. Next, draw a very light weight line from each end of this vertical line to each of the vanishing points on the horizon line (four lines in all).

3. Along the bottom edges, (below the horizon line) visually judge approximately where two additional vertical lines should be placed parallel to the first, which create the proper proportions for the building, and place a point on each of the two bottom edges.

4. From these two points along the bottom edge of your building, draw light construction lines to the upper or third vanishing.

5. Darken in the edges of the prism, judging which lines will be visible, and where the corners intersect.

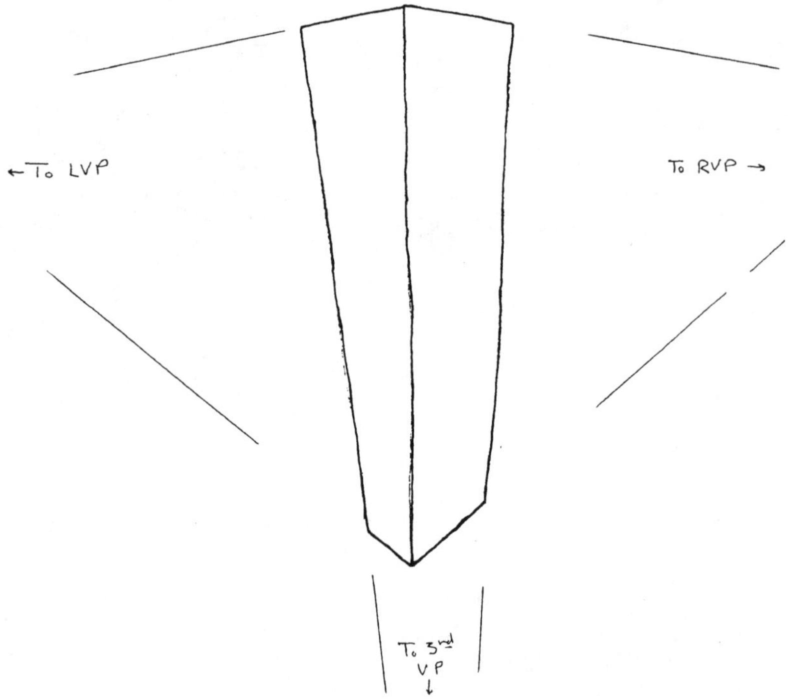

Three Point Perspective

One Point Perspective One point perspective is used to depict the interior of a long hall or room, or, it can be used for other long objects viewed from the interior. To make a one point study, begin with a rectangular prism approximating the proportions of a room or hall way interior. Make sure the prism fills most of the page, or use the edges of the page for this purpose. Select a horizon line below center is you wish the view to be from the eye of a standing person or above center for a sitting person. Place the vanishing point anywhere along this line. Where you place it is determined by the placement of the viewer, i.e. in the center or closer to one wall or the other. Draw fine construction lines from each of the corners of the prism (page) to the vanishing point. At some place along one of these lines, place a point representing the far end of the room/hall. Using this new point as a reference, draw a line through the point parallel to the large prism and successive lines to create a new prism parallel to the first. This new prism represents a wall at the end of the hall. Add details as desired to give the study the appearance of an actual room.

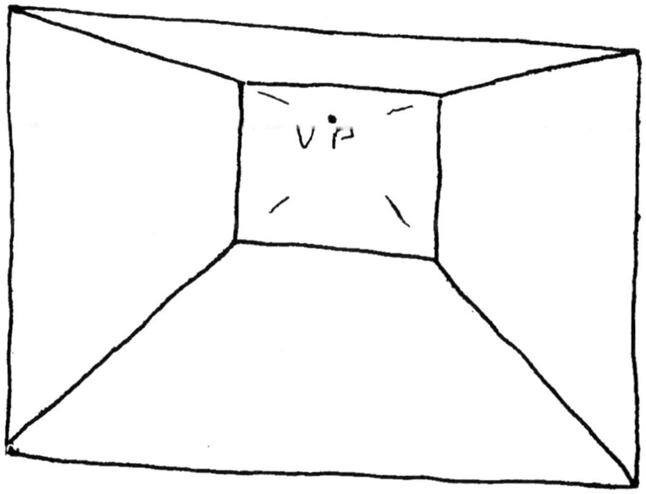

One Point Perspective

Chapter 7 - Light and Value

One aspect of seeing that has been ignored in these studies thus far is the effect that light has on the appearance of objects. Without light we do not see; contour studies, negative space, and perspective studies ignore that basic fact. Light does not reflect evenly off surfaces for a variety of reasons. There are different colors, reflecting differing wavelengths of light, the surfaces are at different angles to the light source, therefore reflecting differing amounts of light toward the viewer, and the source of light varies. Any one or all of these factors may be contributing to the amount of light reflected from a surface. The amount of light reflecting off of a surface is referred to as value in art and the graphics industry. The varying light value of an object adds considerably to its appearance of 3 dimensionality. Accurately capturing the varying light value of the surfaces of an object involves learning to see variations in light, then using one of several methods to achieve value variations on paper, a technique known as shading.

Value Scales

To begin with, practice with some of the following techniques for creating different values. When using a pencil, value variations are created by rubbing more or less of the pencil graphite off onto the paper. This can be done in a variety of ways, but the most common is known as smudging. To get a feel for these techniques, begin with a value chart. Use the following directions.

1. Turn the paper you are using for this purpose and make 5 rectangles on the page. Make the rectangles long and narrow and orient the long dimension from left to right across the sheet. Fill most of the

page with the prisms, but leave a definite space between them. (see illustration, page 54)

2. Divide all but one of the prisms into five to seven smaller, approximately equal boxes. Each the larger rectangles will be used to practice a different technique. Each of the smaller boxes will represent a different value grade. The eye is capable of discriminating about 12 to 15 value variations, but it is difficult to create more than about nine different values without a great deal of practice.

3. In the first series of boxes, the value differences will be created with the smudge technique. A plain, number 2 office pencil works best. Lay the pencil sideways such that the entire exposed portion of the lead is in contact with the paper. Rub the lead against the paper to darken in the box. The first box will represent the darkest value obtainable with this technique, and the last box will represent the lightest value by leaving it blank and allowing the white of the paper to be exposed.

4. Fill in the first box by rubbing off lead until it is very dark. In each successive box, rub off less lead so that there is a definite change in value. Fill all but the final box in this manner creating a series of value differences from darkest to lightest.

5. In the beginning, it may help to begin at the light end rubbing off only a small amount of lead and then repeating this in the remaining boxes. By repeating this technique for each box, value differences are easier to achieve.

There are numerous other ways to achieve value differences. We will review three more of these here. A technique often used by graphic artists, technical illustrators, and industrial designers is that of varying line density. In the second row of boxes, use the following technique:

1. In the first box, draw a series of vertical lines as close together as you can get them. This first box, as in the row above, represents the darkest value. One advantage of lines is that they can be made with a

variety of media, including ink. One disadvantage is that they take longer.

2. In the next box, make a series of lines with a noticeable space between them. In each successive box, increase the space between the lines, leaving the last space blank to represent the lightest possible value, the white of the paper.

When viewed from a distance, each box represents a different value, governed by the amount of white from the paper showing through the lines. The eye tends to blend the lines and white and see different values. The technique is not particularly useful for organic shapes, but works quite well for geometric forms.

The next technique you will try is that of dappling. This technique does not work very well with a pencil unless it is a special artist's pencil with a very soft and large lead. It is suited best for brush-tipped pens or markers, and you may want to experiment with one of these media. The technique is as follows:

1. In the first box, tap the point of the drawing instrument against the paper causing dots to appear. Continue this until the first box is almost completely filled with dots or dapples.

2. In the successive boxes, dapple the paper somewhat less, leaving successively more of the paper showing through, decreasing the apparent value. Again, from a distance, the eye tends to blend the dots or dapples with the white of the paper to see different values.

The last technique, crosshatching, is one commonly used by artists when drawing with ink. Unlike the graphite from a pencil, it is rather difficult to control the amount of ink laid down on the paper. The technique is similar to lines, but much more adaptable to organic form. This technique is accomplished as follows:

1. In the first box of the next row, make a series of lines in the box which are diagonal. They can be made quickly and do not need to touch.

Make another series of lines in the same box, also diagonal, but at a different angle, which causes the new lines to cross with the old at approximately right angles. The angle between the lines is not really important.

2. Continue to add more lines at different angles until the space in the box is nearly filled. In the next box, begin with the first series of lines, slightly farther apart. Add additional lines but not as many as in the first box, causing the value to be slightly lighter.

3. Continue the process in each successive box, using fewer lines, spaced further apart causing each box to have a lighter value than the one before it, leaving the last box blank. As in the line and dapple techniques, at a distance, the eye tends to blend the hatching and the white of the paper to perceive variations in value.

The last rectangle drawn on this practice sheet should not have been divided into discrete boxes. Value in objects does not always have sharp divisions. Value differences may be sharp, as in cast shadows, but not always, particularly with light reflected from organic shapes or cloth objects. In the last rectangle, select one of the techniques used in the discrete boxes, and create continuous tone value changes from the darkest at the left end of the rectangle to the white of the paper at the right hand end. Try not to create any sharp changes in value which appear as lines in the box. A handy tool for this purpose which can be used in conjunction with a soft pencil when using the smudge technique is a paper stick. This is a rectangular piece of heavy paper wound into a tight spiral such that it tapers to a point at one end. The stick can be used to blend the graphite on the paper to remove sharp contrasts in value in adjoining spaces. Use any technique you feel comfortable with, but practice all of the techniques. It will take a lot of practice to master these techniques. Experiment with all of them to find the one or two which work best for you. To begin this practice, complete the series of value studies described in the next section.

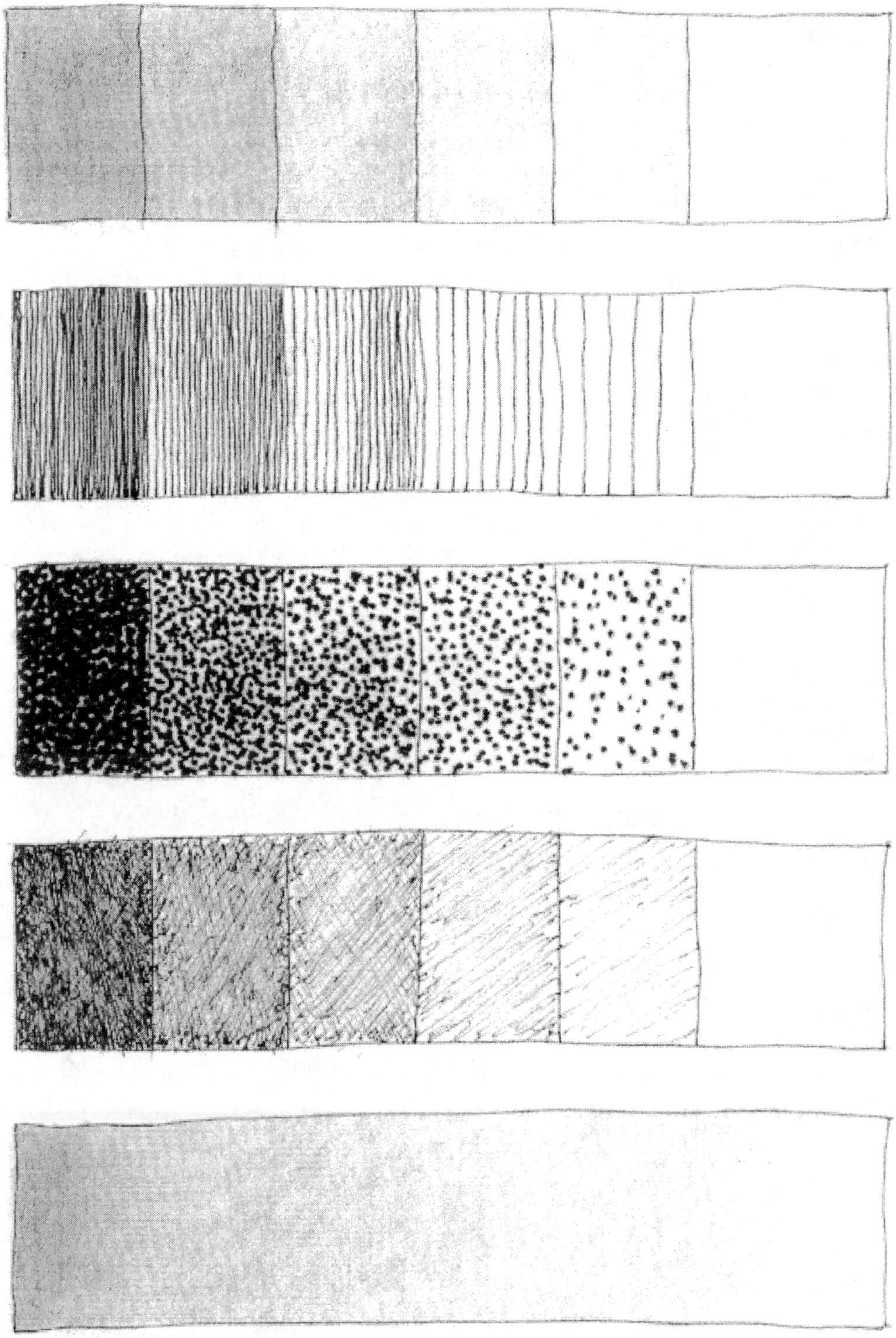

Value Scales

Value Studies

There are two ways to approach a value study. You may do a modified contour study first and then add value (this process is referred to as rendering), or you may do a pure value study, where the shape and form of the object are described by variations of value. The second of these techniques is by far the most difficult and although you should try it, it is suggested that you begin with a rendered study.

Lighting is very important when doing value studies. Diffused fluorescent light, such as you might find in a classroom, does not create a high degree of value contrast, making the value differences difficult to see. The best lighting condition is with a lower wattage lamp to one side of the object being studied.

Start by making a modified contour study of the hand. Do a new one rather than using a previous one. You can use the practice. Once you have completed the contour study, take another look at your hand. This time focus on the value differences, where it is dark and where it is light. Notice how many discrete levels of value are present and where they are. Using whatever technique you are most comfortable with, duplicate the values that you see into the contour study. Begin with the darkest values and work toward the lightest. Remember to use the same approach to these studies as described earlier. You want to be able to make the right-mode shift into visual processing.

When you have completed this study, take a look at the results and compare them to your hand. Most likely you will not be particularly satisfied. That is normal unless you possess an exceptionally high degree of artistic talent which most of us don't. It is not normal to expect perfect results on the first try, but as in most things, you will improve with practice. Repeat the study and concentrate on seeing and replicating the value differences.

Two additional and very challenging value studies which will improve your abilities more than most, are white object arrangements and cloth object studies. Before you begin with either of these, it is recommended that you do a study of a white ball. The foam spheres that are found in most crafts stores work well for this purpose. Set the sphere on a table with a light surface, and place a table or reading lamp close by. Overhead lights should be turned off for best results. When you look at the sphere under these conditions, you will notice a light area nearest the table surface, then a dark area, then lighter with a highlight on the surface of the sphere nearest the light source. Draw a circle on your paper representing the contour of the sphere. Draw additional very light lines outlining where the different value areas appear. This technique is called value mapping and can be very useful in learning to see value differences in objects, particularly when those differences are very subtle. After you have "mapped" the value area of the sphere, fill in those areas using one of the rendering techniques you have just learned, until you get the appropriate value in each area.

After you have completed the sphere, collect several additional white objects, such as a tissue box, golf ball, a Styrofoam cup or whatever is available. Place them in an interesting arrangement or stack on the table with the light source again to one side. Complete a modified contour study of the arrangement first, then add value as you see it.

For your next study, select a piece of clothing or possibly a towel. An unmade bed will also do for this study. In this study, work at defining the contours in the cloth as differences in value. These areas of high value contrast should be very apparent, although this study will be one of the most challenging. As in all skills, and particularly those with which you are unfamiliar or have no previous experience, you will need to spend a lot of time practicing. You should work with additional value studies for several hours before going on.

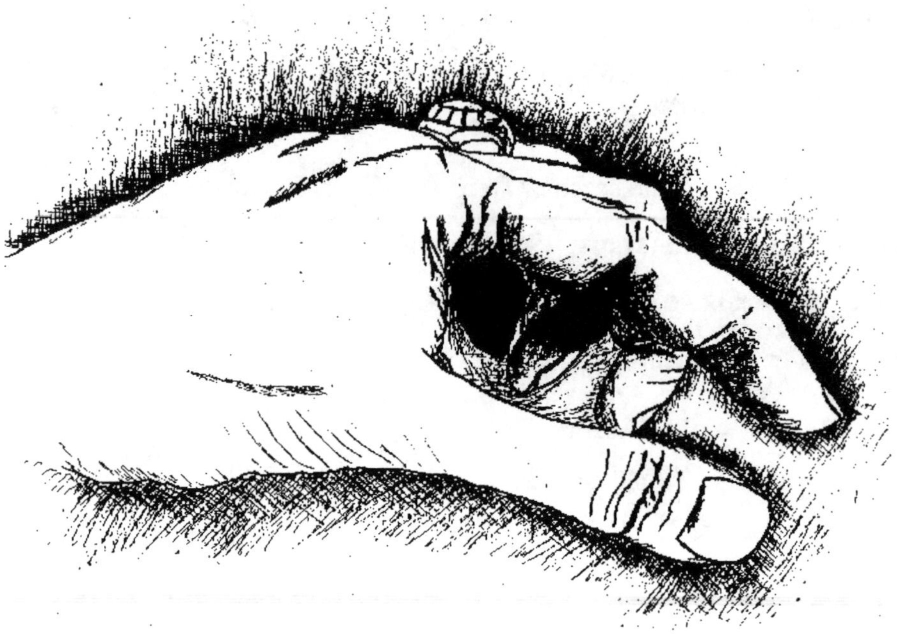

Value Study of a Hand

Chapter 8 - Color

As stated in the previous chapter, light is required for the eye to perceive shape and form. The white light of the sun, the purest source of light falling on this planet, contains all of the different colors in the spectrum. We know that color is the result of different materials absorbing some wavelengths of light and reflecting others. There are several theories of color from science, art, photography, the printing industry and more recently, television and computer monitors. From science, we know that pulsing magnetic fields radiate energy through space which is how we get light from the sun. Radio and television transmission take place using this same principle. The varying frequencies of pulsation of the magnetic radiation create different properties. Within a particular range of frequencies, the radiation turns into visible light. At the lower end of this range is red light, and even lower, infra-red, and at the upper end of the range is violet or even higher, results in ultra-violet radiation. All of the colors of the rainbow exist in between. A rainbow results when the light passes through rain droplets, which act as prisms, resulting in the diffusion of light into all of the frequencies in the light spectrum. The colors that we can differentiate clearly are called hues and consist of red, orange, yellow, green, blue, and violet, also known as the primary colors, or those contained in a color wheel. The colors in this wheel can be mixed to achieve a wide variety of hues, but if opposites on the wheel get mixed in equal amounts, the result is neutral gray. Even though the combinations are numerous, hue alone does not give us all color possibilities.

There are actually three dimensions of color: hue, saturation, and value. Hue we have just discussed, although there are variations of this theory in art, graphics, and monitors. Which ever theory is used, hue is the purest form of the color, and is either one of the six primary colors or a mixture of adjacent or opposite colors. Saturation is the dimension of color which defines the degree of purity of the hue. When opposites get mixed, then saturation is diminished. If all of the primary colors or two opposites on the wheel get mixed, the result is called neutral gray. Differing amounts of opposite colors, such as red and green, result in differing degrees of saturation, giving each color a grayish cast. Value in color is the same as value in shading, and is a measure of the light intensity reflected from a surface. An illusion of the amount of light can be created on paper with

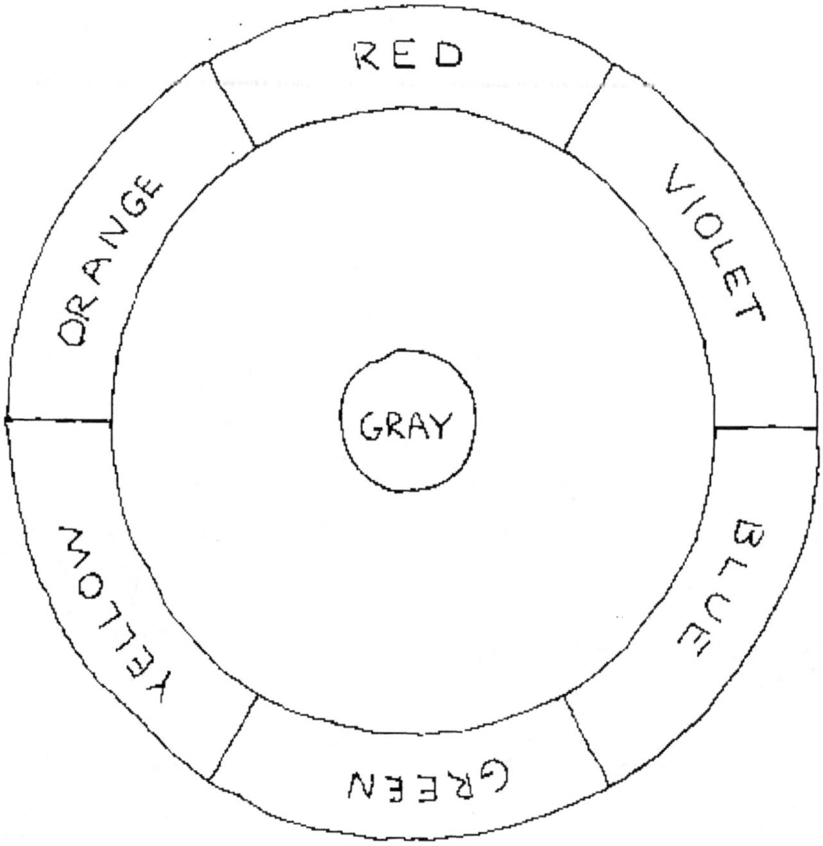

Color Wheel

Opposites

Orange	**Blue**
Yellow	**Violet**
Green	**Red**
Cool	**Warm**
Advancing	**Receding**

colors the same way it can with gray scale shading, by adding black or white to the hue. Hues with white added are called tints (pastels) and those with black added are called shades. To demonstrate this concept, we might choose a familiar color such as red. Add white and we get pink, a tint, and add black and we get maroon, a shade. With the six primary colors, mixing to get variations of hue and differing degrees of saturation, and by adding light (white) or subtracting light (adding black), all of the color variations that the eye can distinguish (and more) can be achieved. With a computer, these variations can number into the millions.

With sunlight and electromagnetic radiation the primary colors are red, blue, yellow, green, orange and violet. With ink in the printing industry, cyan, yellow, magenta and black with the white of the paper achieve all of the colors that you see on the printed page. In television and color computer monitors there are three guns shooting varying amounts of electrons which light up red, green and blue dots (RGB) on the screen at differing intensities. The mixture of these dots and the black in between create all of the colors. White is achieved when all three guns hit the same place with the maximum degree of intensity. Pigments, paint, and other artist colors do not always get results according to these theories, because not all pigments are pure hues. When mixing some pigments, red, blue, and yellow result in brown rather than neutral gray.

It is not the intent of this book to address studies in color. This discussion was intended to address some of the more frequently asked questions about the nature of color. If you wish to conduct some simple experiments, the easiest to use colors are chalk pastels. A set of 10 or 12 is sufficient and not very expensive. Create a color wheel with six spaces and use the red, blue, and yellow pastels to fill in every other space. Use equal amounts of red and blue and you should get a shade of violet; red and yellow result in orange, and blue and yellow result in green. Green will most likely be the least successful in this experiment, probably having something to do with the purity of either blue or yellow or both. Another visual principle of colors is that they are either advancing or receding. Red, blue, and violet are recessive colors and yellow, orange, and green are advancing colors. To demonstrate this effect, draw three large squares with a smaller square inside of each. Fill in the outside square with an advancing color, and use its opposite on the color wheel, which will be a receding color, to fill in the inner square. For example, blue and orange are opposites on the color wheel and work well to demonstrate this principle. Color in the outer square with orange and the inner one with blue. Notice that the blue square seems to "sink" into the orange square. Try this with each set of opposites; yellow and violet, and red and green.

You can use these same opposite color sets to demonstrate the dimension of saturation. Draw three rectangles on a piece of paper similar to those used for testing the techniques to achieve value differences. Do not fill in the discrete boxes as in the value experiment, however. Begin at one end of a rectangle with any color, and in a manner similar to the continuous smudge technique, make one end very dark with the color and then less and less to about three fourths the way across the rectangle, leaving the rest blank. Use the opposite of that color, i.e., if starting with blue, use orange for the second color, and beginning at the opposite end of the rectangle, do the same thing. The result should be the pure hue of either blue or orange at either end, gray in the center where equal amounts of each color exist and some shades of grayish orange toward the orange end and grayish blue toward the blue end.

You can conduct a similar experiment for color values, by using a hue, and mixing white at one end and black at the other. With the six primary colors and the different hues that can be achieved by mixing them, the various degrees of saturation, and the varying values, the different color possibilities are almost infinite. If you have access to a computer graphics program which allows you to mix colors, you can see how the effect that each color dimension has on the result. Some high end computer graphics programs advertise the capability of 3.5 million different colors. They all are variations of red, green, and blue, mixed in ways varying the three dimensions of color.

If you desire to try a color study, use a piece of fresh fruit or fresh flower for the object. Natural objects have more variation in color and are much more interesting than manufactured products.

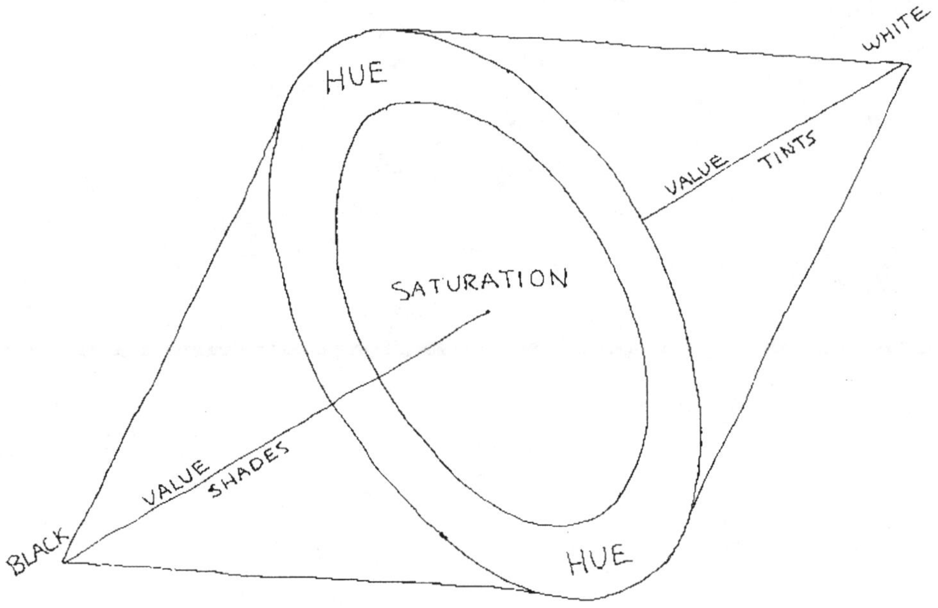

Cones Representing the Three Dimensions of Color.

The Color Cones:

The illustration above represents the relationship between the dimensions of color. The ring at the common base of the adjoining cones represents hues. As hues move toward the center of the base, they diminish in saturation and take on gray tones. As colors move in either direction toward the apex of the cones, they change in value, becoming either darker or lighter, depending on the direction. The axes of the two cones, joining the two apexes, represents the gray scale. Theoretically, all possible color combinations fall within the boundaries of the two cones.

Chapter 9 - Geometric Solids

For some time engineers, architects, and designers working for large corporations have been using 3-D computer aided design/drafting systems in the designing of parts, products, and buildings. However, until just recently, these systems were very expensive, and some are still relatively so. The 3-D modeling system that The Boeing Company assembled to design the 777 jet airplane cost in excess of $12 million. Although not all systems are that expensive, until recently, similar systems would cost from ½ to one million dollars. The advent of powerful Unix workstations have brought this capability to much smaller companies with systems in the $75,000 to $100,000 range, although these can cost considerably more, depending on the use to be made of them. Still, this price range is a fraction of the cost of similar systems of 10 to 15 years ago. Recent developments in the power of micro computers, along with the increase in power of 3-D modeling software, are now making these systems available to the small manufacturer for under $10,000. There is considerable evidence that the world of computer aided design/drafting is making a rapid shift from 2-D drafting to the 3-D modeling environment. There are many different 3-D modeling programs. Some of these programs use 3-D solids and Boolean operations, i.e., solid geometric shapes such as cubes, cones, cylinders, prisms, pyramids, Torii, and spheres which can be combined to form more complex shapes. Most will also have a number of ways to create forms which do not conform to the parameters of regular solids. These are commonly known as boundary models. Others use a process known as feature-based modeling which begins with a basic solid such as a cube. The basic form will have features added or subtracted until the final intent is achieved.

There are many different programs, but for the operator, they all have at least one thing in common, and that is necessity to learn to think 3-dimensionally. In order to become skilled at using and visualizing 3-D models/solids, it is useful to study these shapes in their natural form from observation, recording the process through drawing. It is also useful to follow up the observation studies with mental imaging studies, also recorded through drawings.

Solid model studies

The common geometric solids included in most 3-D Computer Aided Design and Drafting (CADD) modeling programs are cubes, cylinders, cones, spheres, and prisms. The prisms are primarily rectangular prisms although any regular polygon may be extruded into a prism. It is not the intent of this text to explain any of the methods used by the various software programs, except to make the point that each method makes use of geometric solids in some way. The way parts are generated frequently involves combining solids through addition or subtraction, or through generation of non-regular solid form. In any case, the understanding of geometric solids is a valuable asset in designing whether it be done manually or with the use of a 3-D CADD system.

Solid model studies can be very useful for this purpose. Begin these studies with the cube, and begin each cube by first drawing the edge contour closest to you. You can then draw the bottom and top contours at any angle to horizontal, that is, either slanting up or slanting down. If you observe even very small cubes closely enough, you will be able to detect some degree of perspective convergence. However, for the sake of ease in these studies, it is important that the top and bottom edges of the same side of the cube remain parallel. These first "studies" can be from memory, but it is important that at least some be from observation.

Gather up several objects that are cubic in shape, such as a tissue or other box, and use these for at least two or more of the studies. Fill up the page, or several pages, with cubes, varying their orientation (as seen from above or below the cube) and size. In the studies from memory, think of all of the cubes making up a creative holistic design. This will make the study more interesting. When you have finished with cubes, complete similar studies of spheres, rectangular prisms, cylinders, and cones.

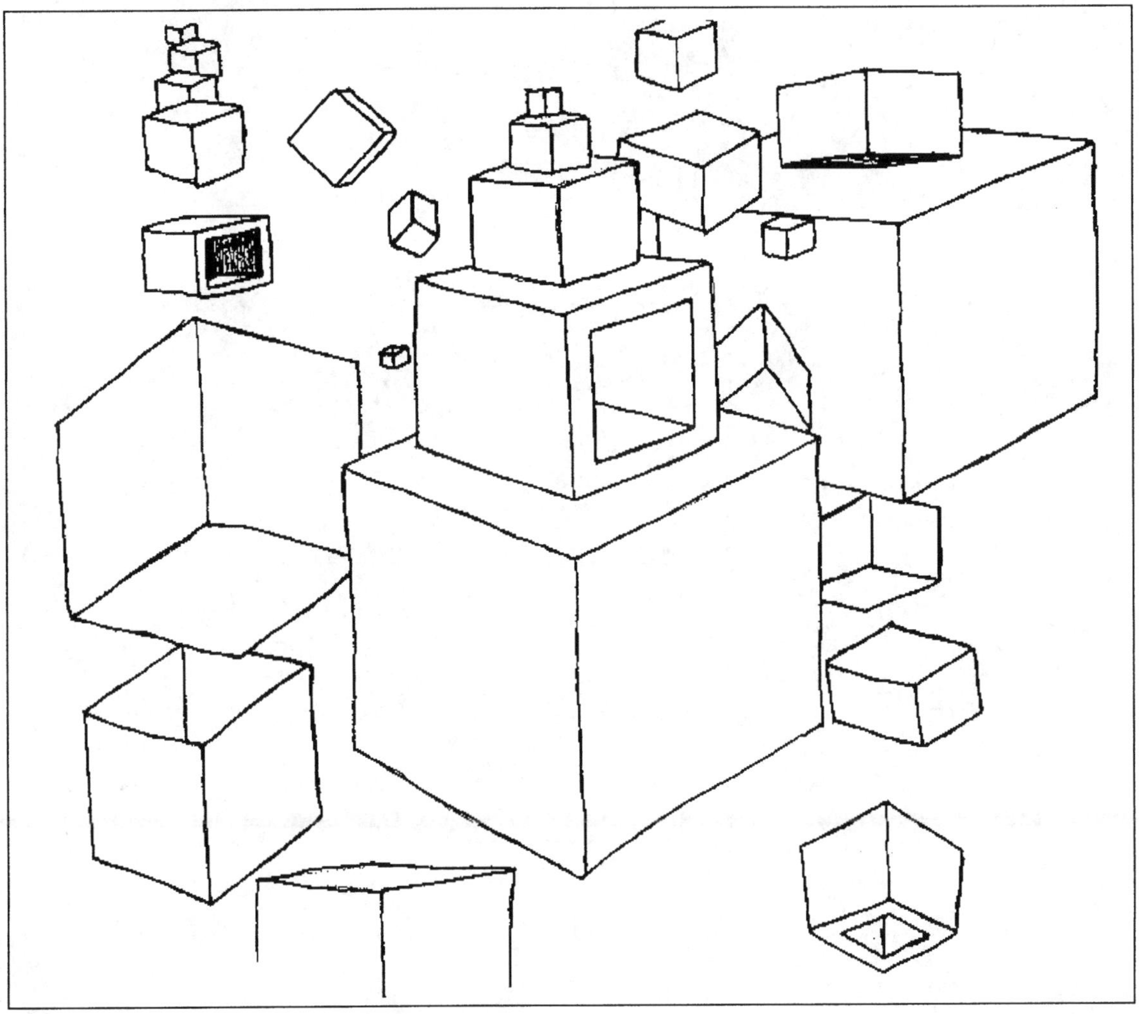

Study of Cubes

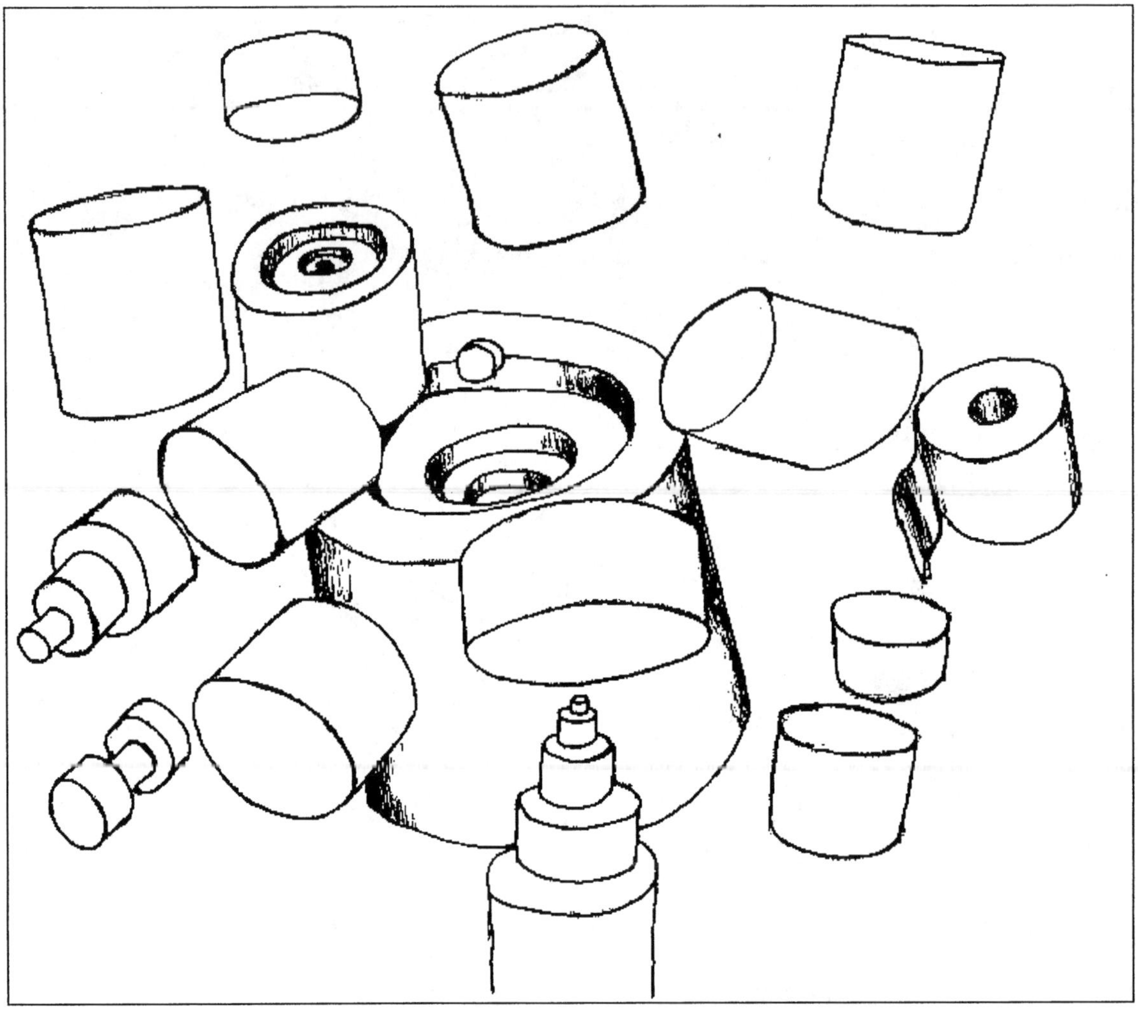

Study of Cylinders

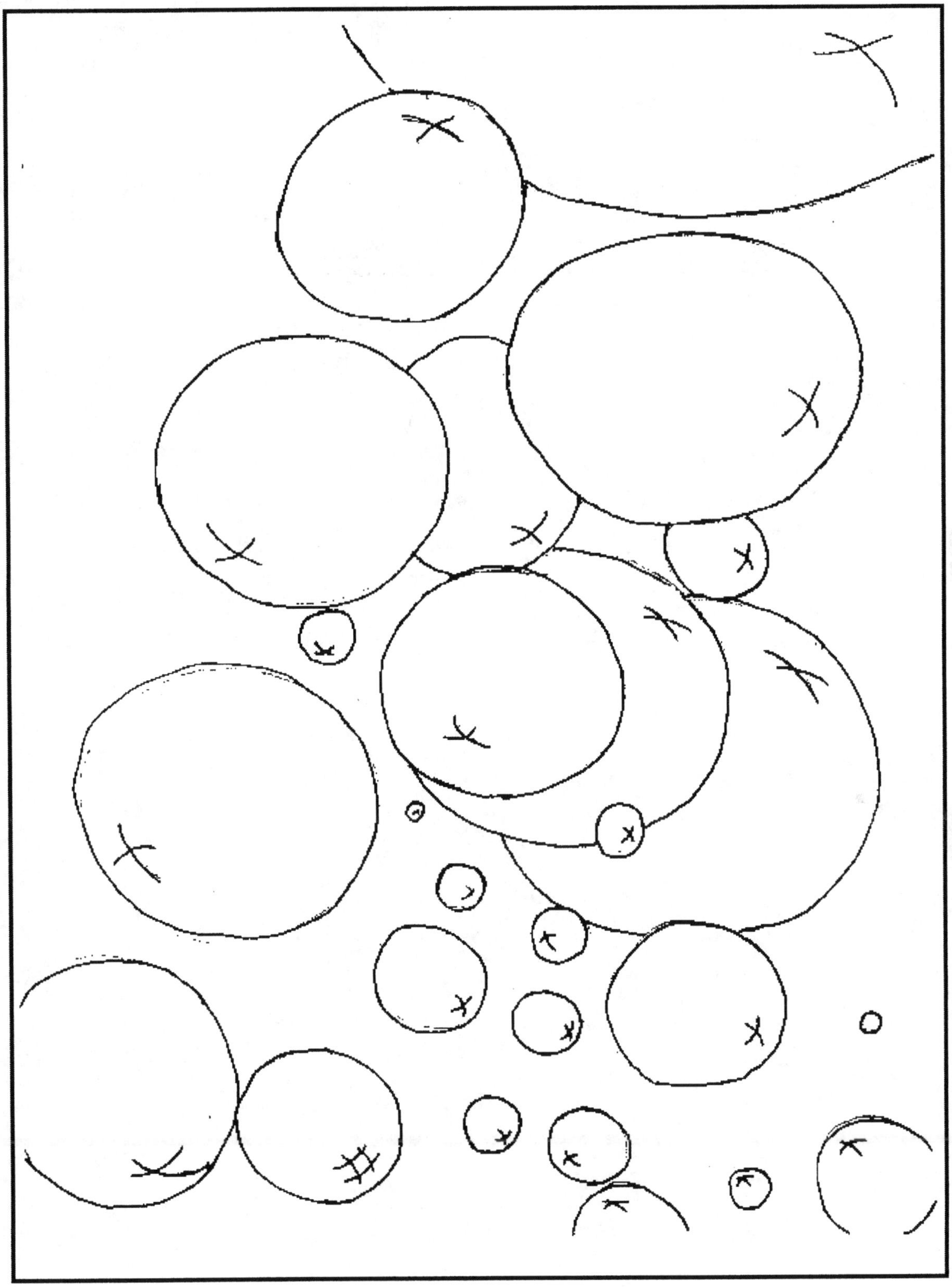

Study of Spheres

Chapter 10 - Axonometric Views (Parallel Projection)

Axonometric view is a technical term used to define specifically how an object is rotated and then drawn from observation. The term is part of the system of descriptive geometry and orthographic projection that will be discussed in more detail in the chapter on Technical Sketching. A brief discussion here is useful, because anonometric views can be helpful in letting the left brain in a little bit on this processing. Something that should be kept in mind, is that true creativity involves the ability to process information in both the right and left modes with equal agility. Axonometric gives us a consistent way of viewing the object and structuring the drawing.

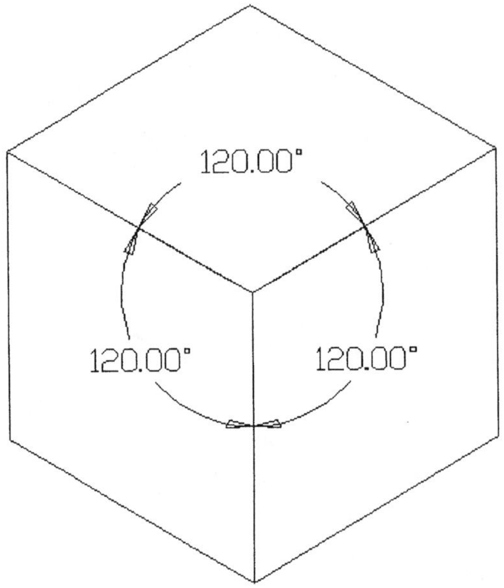

Isometric Cube

Isometric

Isometric views are the most common of the group of axonometric views, mainly because they are the most convenient to draw. The term isometric means "equal measure" and refers to the angles made between the contour edges of the top and two side surfaces of the object at the point of convergence. This relationship is most easily observed in an isometric sketch of a cube. To do this, you must start with a viewing position at right angles to one of the faces of the cube. In this position the cube will appear to be a square with no other surface visible. You then rotate the cube 45 degrees to one side and tilt it toward you an angle of 35.264 degrees. The odd angle of tilt, is what produces an isometric view. If you then focus on the corner of the cube which is closest to you, you can observe that the angles of the three visible corners are all equal. This also results in the corner closest to you and the corner furthest away aligning visually, which cannot be observed unless the cube is transparent. One last condition which may be observed in an isometric sketch. If you were to draw a horizontal line through the intersection of the three boundary edge contours, that line and the edge contours defining the top surface are equal and 30 degrees on either side of the intersection. This makes isometric drawings easy to construct with a 30/60 right triangle, and isometric grids easy to follow.

Practice making isometric sketches of each of the solids, except spheres, from the last sketching exercise. For the cone and cylinder, you may want to start with a square prism to "box" the shape first. You can

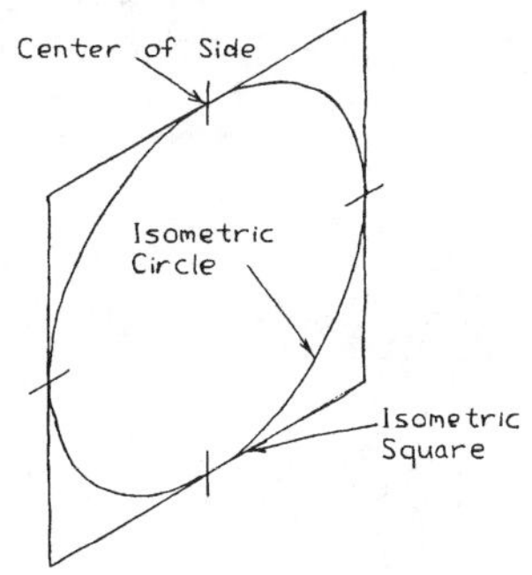

then sketch an isometric circle in each of the end squares (for the cylinder; a dot in the middle of one end may be used for a cone) and then connect the circles at the edges of the cylinder. The technique illustrated below will serve for making isometric circles. It is based on the geometric principles associated with inscribing a circle in a square. Such a circle is tangent to the sides of the square at the mid points. In an isometric square, the circle follows the same principle, resulting in an ellipse.

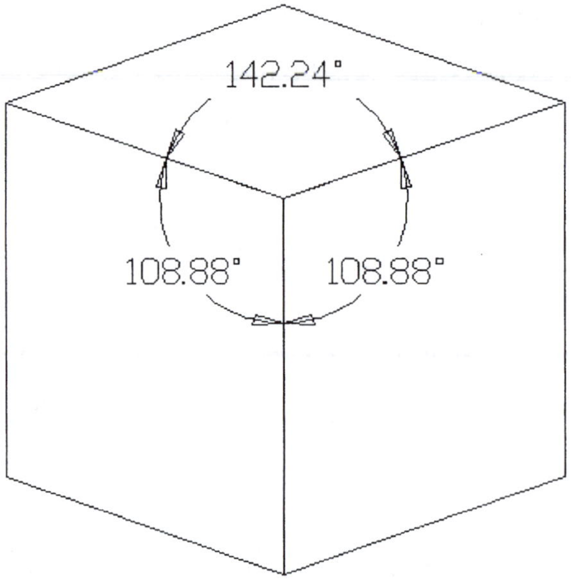

Dimetric Cube

Dimetric drawings occur when only two of the angles between the corners (or axis) of the cube are equal and trimetric drawings occur when none of the angles in the corner of the cube (axis) are equal. These views are useful to show variation in orientation (rotation) of the cube or prism. The dimetric view is obtained by keeping the rotation of the cube at 45 degrees and changing the tilt. The trimetric view is obtained by changing both the angles of tilt and rotation from that of the isometric view.

Obviously, the trimetric view can have more variation. These terms are part of the formal system and are not particularly useful when drawing objects from observation or when trying to get creative orientations from the imagination. The main point here is that no matter how you orient the object, whether it be a cube or rectangular prism, the edges of a side remain parallel. Practice making additional pages of cube studies, varying their orientation to obtain visual variety.

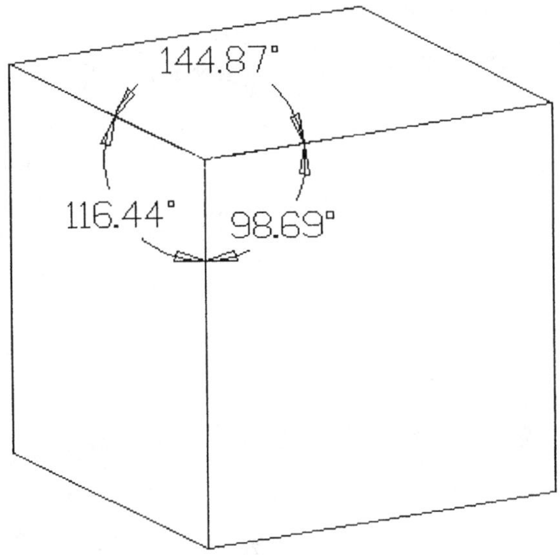

Trimetric Cube

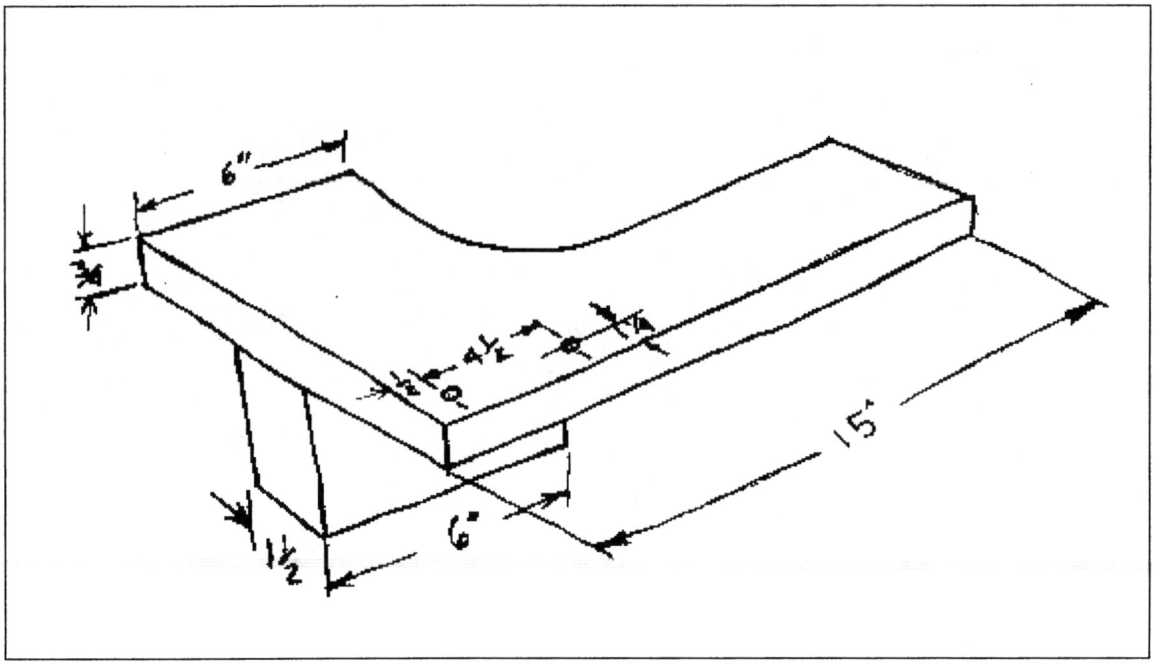

Isometric Drawing

Chapter 11 - Blocking

Although many objects created by humans are basically geometric shapes such as boxes and cans, most of the objects that we design and manufacture or construct that make up our adaptive and/or technological environment are a complex mix of geometric forms. Some objects, such as boat hulls and automobile bodies are organic shapes; that is, they cannot be defined mathematically. Regardless of the shape, the use of geometric form, particularly the rectangular prism, can be most helpful in defining the basic space that an object will occupy. Rectangular prisms can also be combined to make up all of the various sub-shapes that define more complex forms. By stacking and combining basic solids in a drawing, we can define the basic overall shape of a complex object and then through the process of iteration and metamorphosis we can add successive layers of detail until the drawing of the object is complete. Iteration means to repeat and metamorphosis is the process of changing form such as a caterpillar to a butterfly. The iterative process is also known as "blocking out" in the art world. A similar process is used by many 3-D Modeling programs to create parts. The following is an example of how the process works.

The process can be used as an aid to draw complex objects from observation. Its most useful application is to create communications level drawings of original designs. To get a feel for how it works, begin with a complex object and using the iteration technique described here, draw it from observation. Start by studying the object for several minutes as you might do in a modified contour study. Become familiar with the basic geometric solids that make up the object. Use a clean sheet of paper, which is not too heavy, so that lines from a page underneath will show through. Begin with prisms. You could begin by drawing a rectangular prism which encompasses the entire object; i.e., its maximum width, length and depth. A more helpful approach is to create several prisms that bound various portions of the object. Start with a prism representing the core or foundation of the object. Next, attach any additional prisms to the basic one at appropriate places which define the spaces taken up by the details of the object. Many of these parts will be other solid shapes. Eventually, you will convert them from prisms to whatever the desired form is in successive steps. At this point it is more important to focus on appropriate placement of parts and proper proportions. Be careful to use the entire page so that the object will be large enough to show sufficient detail. The result of this first step should be a stack of rectangular prisms which bound the basic form of the object and properly locate and bound the detail forms. If the object is complex including many details, the first step may actually be done with two or more sheets.

Next, tear the sheet you have just completed from the drawing tablet and place it underneath a new sheet. If you are working on a table top rather than in a tablet, make sure to tape the two sheets to the table so that they will not move around. The second step is to convert the prisms to other solid shapes as is appropriate, i.e., cylinders, cones, spheres, etc., using the outline of the prisms underneath to "contain" the new shapes. After this is complete, take the second sheet and place it under a clean one, or place a clean one on top of it. Begin the process of adding detail to the shapes as observed in the object being studied. This process may be repeated as many times as desired to capture all of the detail necessary to complete a highly accurate study. In the early stages, it is not important to have neat, straight lines or accurate curves. This can be accomplished in the final stage.

This process can be very helpful in creating designs from the imagination. It is important to have a basic idea of the form the object being created is to take. In the design process this step is accomplished through simple idea sketches. Idea sketches need only to communicate with the designer, as a form of visual shorthand to record ideas for consideration as the object begins to take form in the imagination. The iteration process is a technique which can aid the designer, particularly in the early stages of skill development, in producing more accurate communications level drawings which can be used to convey the idea to others. This process can also be used to work out the details of a new design. As much detail as desired can be incorporated into these drawings including shading, color, and perspective.

Complete another study of an object created from the imagination. It may be helpful to make a few simple "thumb nail" sketches to work out your idea before you begin the iteration study. Once you have a design in mind, begin with a new sheet of paper and proceed with the observation study. Create a stack of prisms which describe the basic form of the object as well as some of the details. Fine detail is best left to the later iterations. Use the next sheet to convert the prisms to other appropriate forms. These forms may be other geometric solids, or the corners may be "bent" or "curved" to create more organic forms. Use as many different iterations as necessary to work out all of the details of the design.

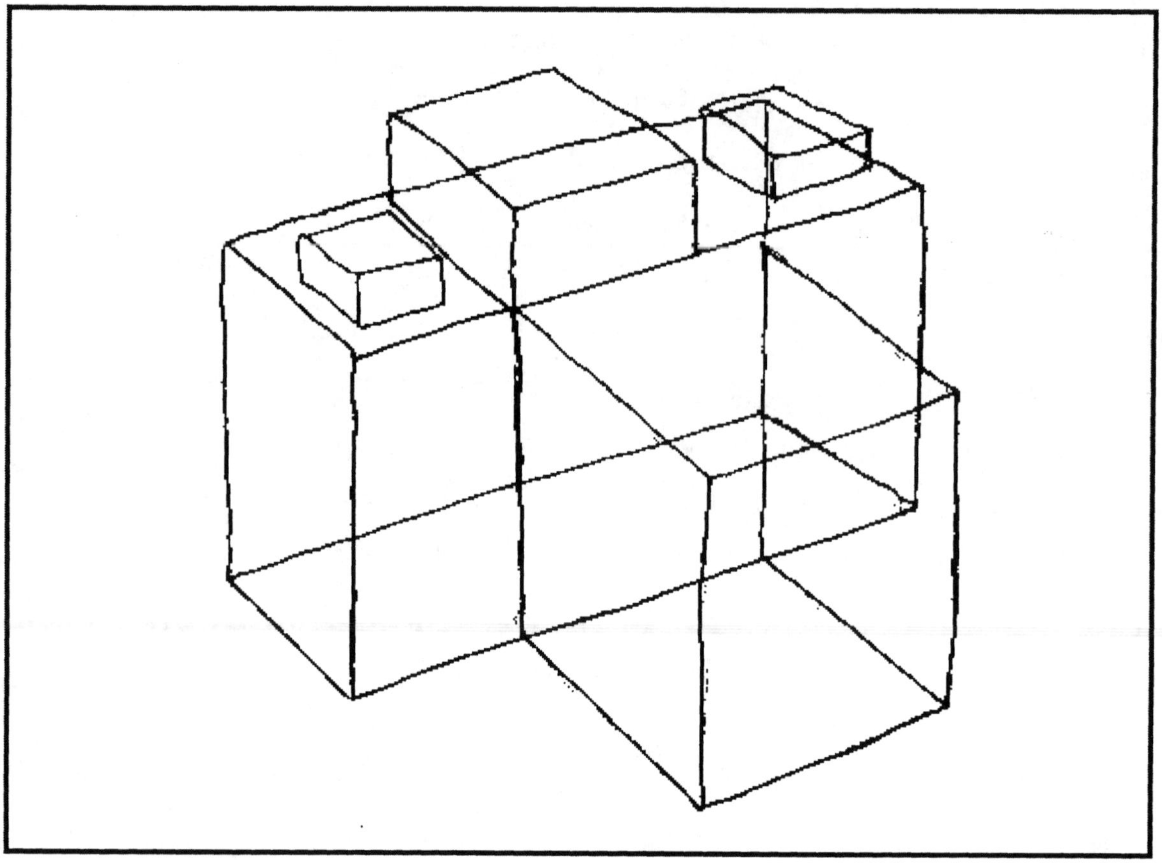

Iteration Number One

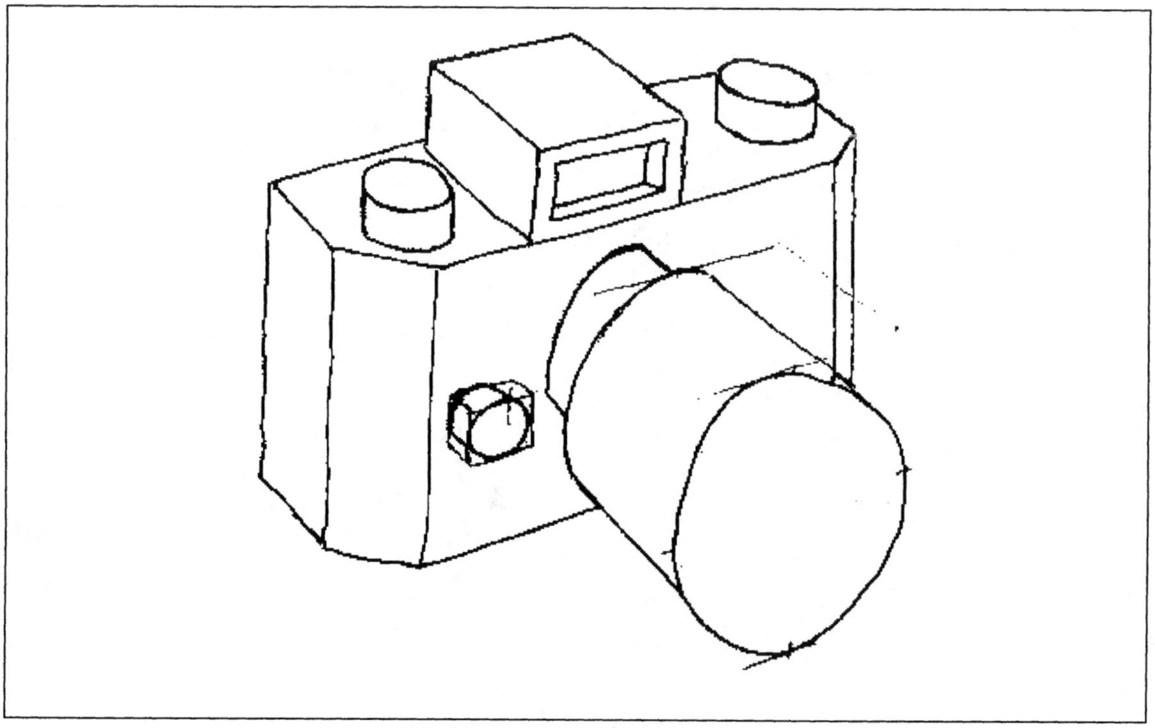

Iteration Number Two

Iteration Number Three

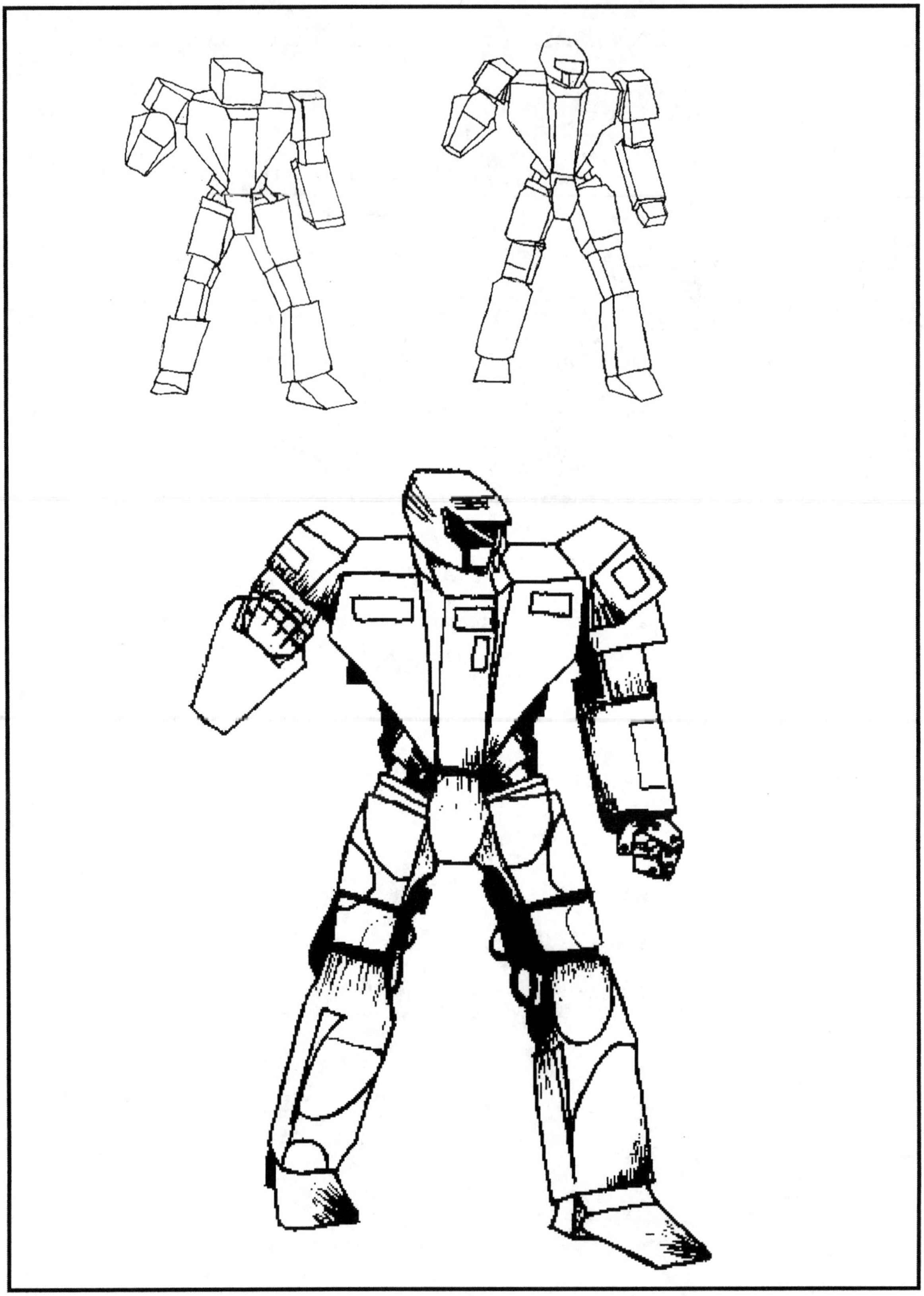

Student Iteration Drawing

Chapter 12 - Technical Sketching (Adding Detail)

There comes a time in the design process when it is necessary to cease improving the design and document the results. This can be somewhat informal to aid in constructing a test prototype, or more formal to guide the manufacturing or construction process. The system used for this purpose was originally developed to allow for the 2-Dimensional description of complex and technical 3-Dimensional objects. The process may be becoming obsolete in some areas of design due to the advent of advanced 3-Dimensional Computer Aided Design packages, which contain engineering analysis and machine programming capability. However, the process is still helpful to work out the details of a design, particularly where multiple parts are involved, before attempting to model it on a computer.

The formal system is know as orthographic projection and descriptive geometry, and was invented in the 18th century. It consists of 6 basic "face" views which represent the surfaces of a cube or rectangular prism. The views all have a right angle relationship like the sides of a cube. It may help to understand how the system works by imagining an object, preferably a human made one, placed inside of a glass cube where it can be viewed. The object must be oriented in the cube such that the surfaces of the object are parallel to the sides of the cube. If we imagine that the glass sides of the cube are also photographic plates, then an image of one side of the object is projected to the glass plate that is adjacent to it. Each side of the object is projected to the adjacent glass plate which makes up the adjacent side of the cube. The glass cube (since we are using our imagination here, each side of the glass cube is connected by a hinge) can be opened flat and will contain an image of one side of the object on each plate.

Describing a 3-Dimensional object 2-Dimensionally requires more than one view, but not all six views. For the sake of brevity and clarity, only the views that are absolutely necessary to reveal the required detail of the part or assembly are used. There are certain conventions for selecting which views to include. Without going in to a lot of detail, there are at least two systems of projection which govern these conventions. In the United States, we currently use what is known at 3rd angle projection. This means that the front, top and right side views are the ones most commonly included. In the case of cylindrical parts, it may be necessary to include only two views which could be the front and either the top or right side, whichever is more appropriate for clear representation and inclusion of dimensions. Occasionally, a fourth view is added if the part or assembly is sufficiently complex that not all of the detail can be shown in three.

Selecting the side of the part to represent the front view also follows two simple conventions. First, if there is a logical orientation to the part from the position of the operator of the machine or object to which it is attached, then that will be the front view. If no such logic is apparent, then the view which reveals the greatest amount of detail should be chosen as the front view. Once that view is determined, then all other views follow logically.

When a part has considerable internal geometry such as complex holes or cavities, it may be necessary to show the inside of the object to fully understand the part. When using a 3-D CADD modeling system, functions are generally available to "open" the part for internal viewing. When using a two dimensional medium, whether it be paper or electronic, it is necessary to create a separate view known as a section view. As in all else that occurs within the standard system of engineering and technical drawings, there are many rules and conventions governing the creation and location of section views. These conventions and standards depend on the type and specific geometry of the part.

There are also many rules and conventions governing the placement of dimensions. Computer programs have been developed to control many of these conventions, however, computers cannot determine which dimensions to include or where to put them.

There are volumes of standards guiding the placement of documentation information on drawings. It is important for the designer to know that there are serious limits to how exact something can be built, and considerable variability in the degree of accuracy with which it needs to be built. While any size or part geometry can be stated or designed with perfect accuracy, and a computer model can be defined with the same degree of accuracy, it is something else again to achieve that accuracy on the shop floor. Machines vary greatly in the degree of accuracy attainable. In the past, the accuracy of parts produced on manually operated machines depended almost exclusively on the skill of the operator. Today, with machines being controlled mainly by computers, much greater precision can be maintained, but at a price. The more accuracy required, the more expensive the process. If each part were designated to the degree of accuracy to which today's most accurate machines are capable, these parts would be too expensive for the market. It is up to the designer to understand the degree of accuracy demanded by the function of the design and to specify accordingly. For example, the accuracy needed for a space craft, or a precision medical instrument, is far greater than that of a wood burning stove. The following is a brief discussion of the different aspects of accuracy that a designer may need to consider.

The size of a part or object is the most common consideration for accuracy. Size accuracy will depend on the nature and intended function of the object being designed, the nature of the materials used, and the availability of accurate machinery and skilled operators. In parts machined from metal, accuracy to four decimal places is common. Precision up to six decimal places is possible with specialized equipment. In parts made from wood, three decimal places is the best that might be hoped for. The desired accuracy is generally designated by stating the limits of fluctuation from the desired dimension and is called tolerance. This can be stated as a limit fraction either above or below, only above or only below the desired size. Other considerations include mated parts and how they are to fit together. The designer may want them to fit loosely, snugly, or so tightly that they have to be forced together so that they hold without added fasteners or devices, such as a wheel bearing pressed into a bearing race.

Another important consideration in assigning tolerances to a part is the positional relationship of the various features of the part. Frequently a part will have a surface or other feature designated as the basis for determining the position of all other pertinent features. In this case, this surface is referred to as the datum. The datum is frequently the reference point for machining the part.

The third aspect of designating the accuracy of a part refers to the geometric shape of the part or the various features which make up the part. How straight and cylindrical holes are depends on the method and the type of tools used to make the hole. For example, machining accurate holes requires two steps; i.e., drilling and reaming. Other considerations include parallelality, angularity, flatness, perpendicularity or any other geometric relationship which might be necessary.

In all cases, necessary accuracy depends on function. The parts in a turbine engine need to be more accurate than those in an internal combustion engine, which in turn need to be considerably more accurate than the wood burning heat stove mentioned earlier. Since cost and accuracy have a direct correlation, more accuracy than absolutely needed within specified safety limits is not economical.

Dimensioning:

There are a number of basic standards for the placement of dimensions that you ought to know. They will assist you in the proper placement of dimensions in both your technical sketches and in CADD documentation drawings. Even though most CADD programs have automatic dimensioning functions, they control only the dimension text, extension lines, and dimension lines. The proportions of arrow size, offset distance, and dimension line extension can be controlled by adjusting dimension settings. CADD programs are not yet smart enough to know what features need dimensioning and where they are to be placed. The following diagram is an example of dimensions placed to appropriate ANSI (American National Standards Institute) standards.

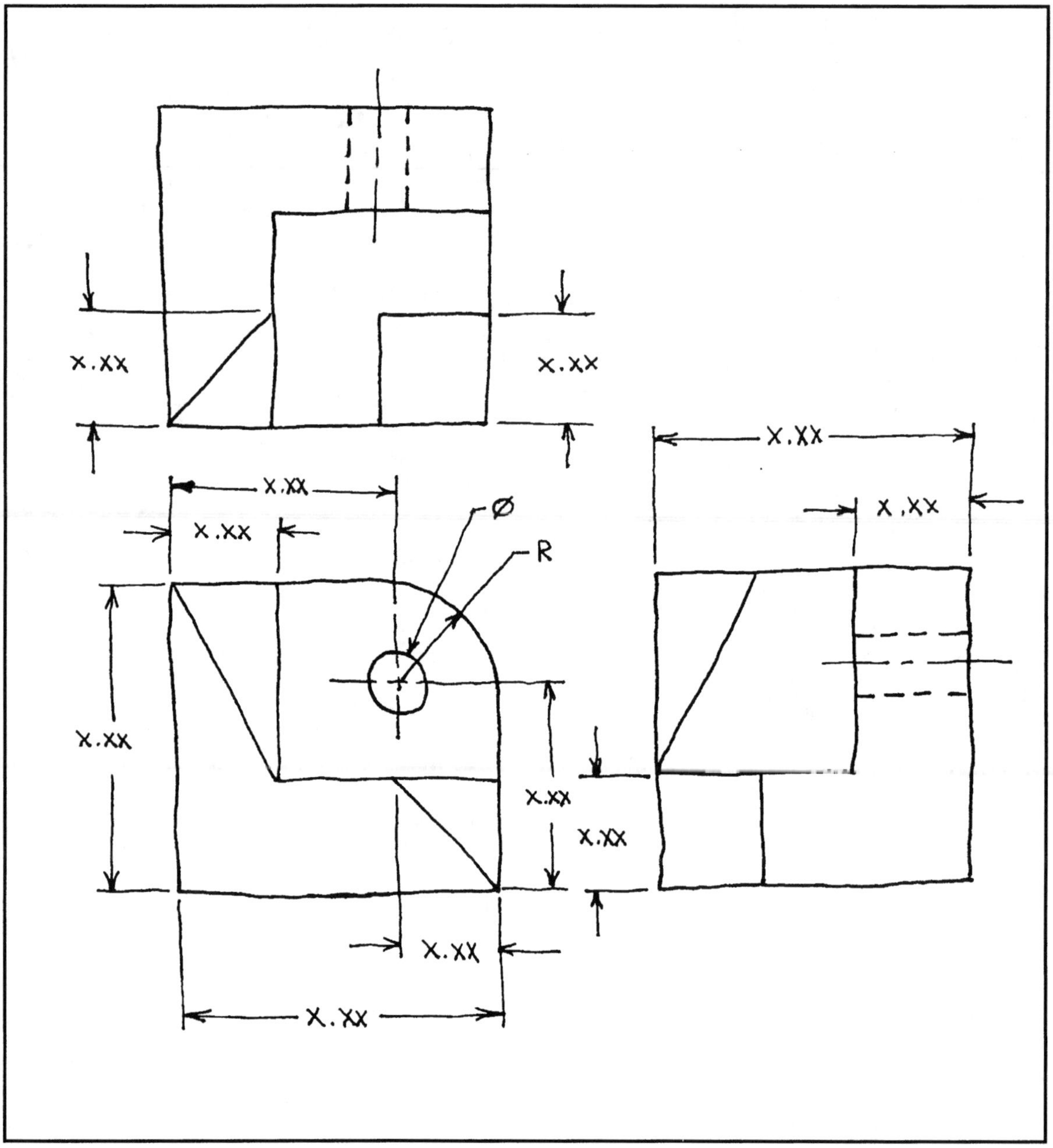

Illustration of Dimensions to ANSI Standards

There are 14 basic standards included in the illustration. This by no means exhausts or even begins to cover all of the standards, but it does include the most common and basic ones. Following these standards will result in technically cleaner technical sketches and initial documentation drawings. Final documentation drawings will most likely require substantially more information.

1. Each feature in the part must be fully dimensioned. (A feature includes things like holes, rounded surfaces, inclines, grooves, notches, etc.)

2. Each feature in the part must be dimensioned only once.

3. Features should be dimensioned in the contour view, or the view which shows the shape of the feature.

4. All dimensions must be placed outside the object.

5. Overall dimensions should be included.

6. Extension should be included which lines track from a feature inside the object and move the dimension to an open area.

7. Dimension lines indicate the extent of the dimension and are indicated by endmarks such as arrows.

8. Shorter dimensions are always placed inside longer ones so that extension lines will not cross dimension lines.

9. There is always an offset (gap) between the object and an extension line.

10. The diameter of a cylindrical feature or part is always dimensioned in the rectangular view.

11. The diameter of holes is always shown in the circular view.

12. The location of holes are always shown in the circular view.

13. Never dimension to a hidden line

14. The radius of an arc is always indicated in the contour (partial circle) view.

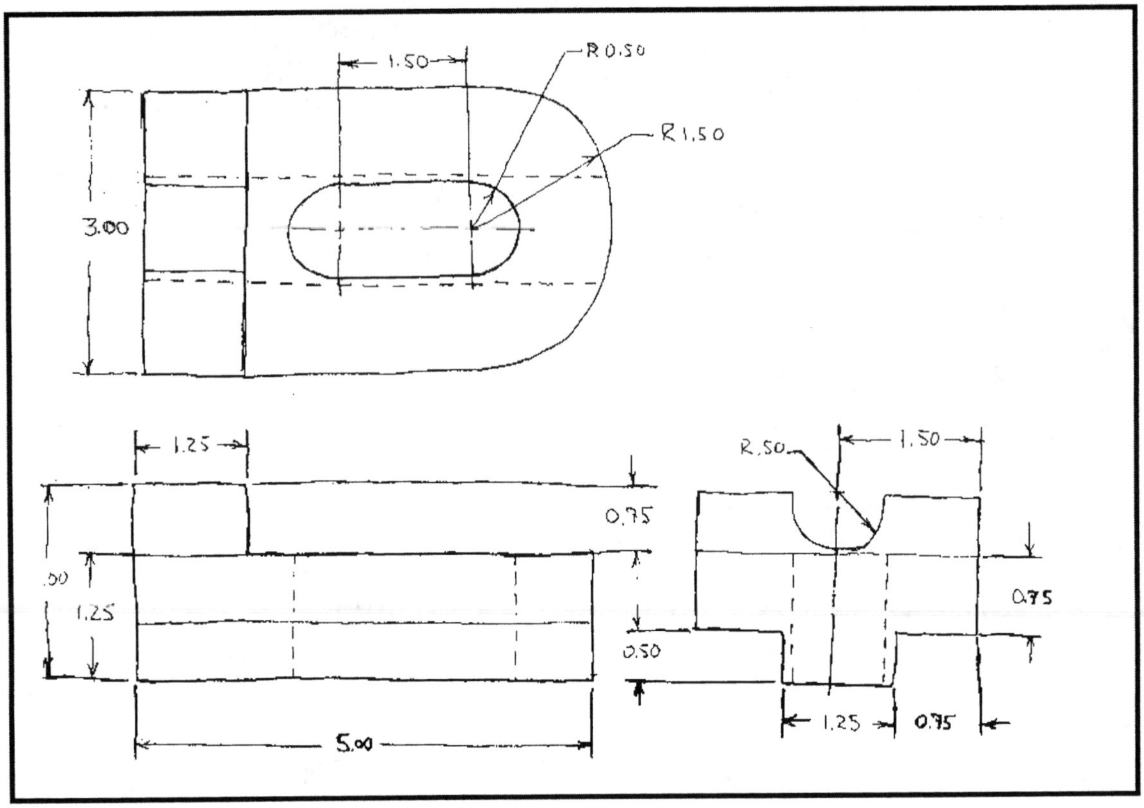

Quick Freehand Multiview Sketch of Block With Dimensions

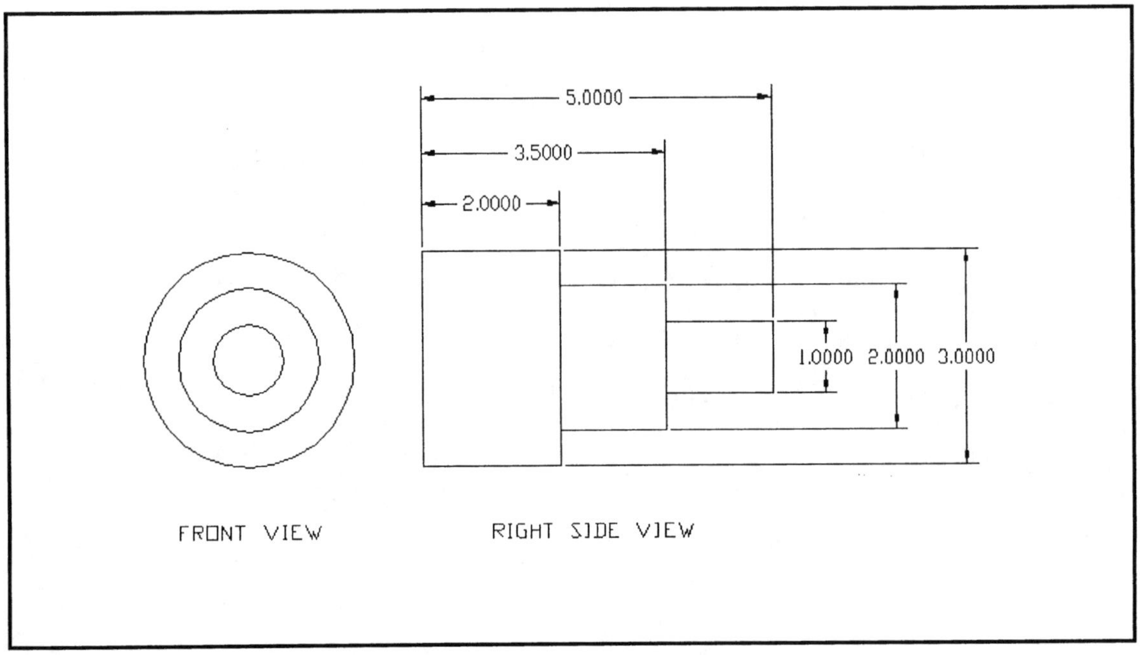

Computer Generated Dimensions

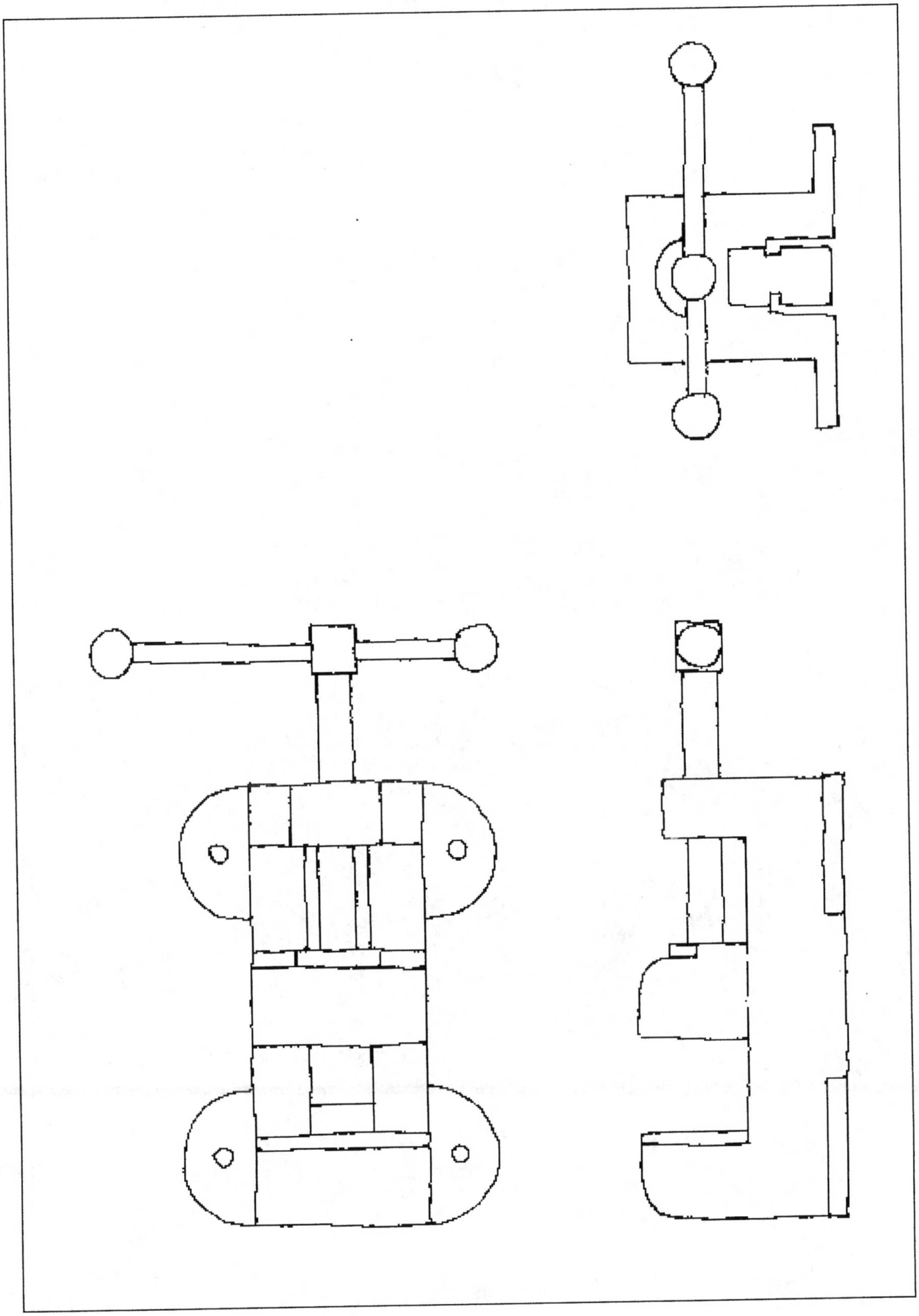

Technical Multiview Sketch (Using Grid Pape

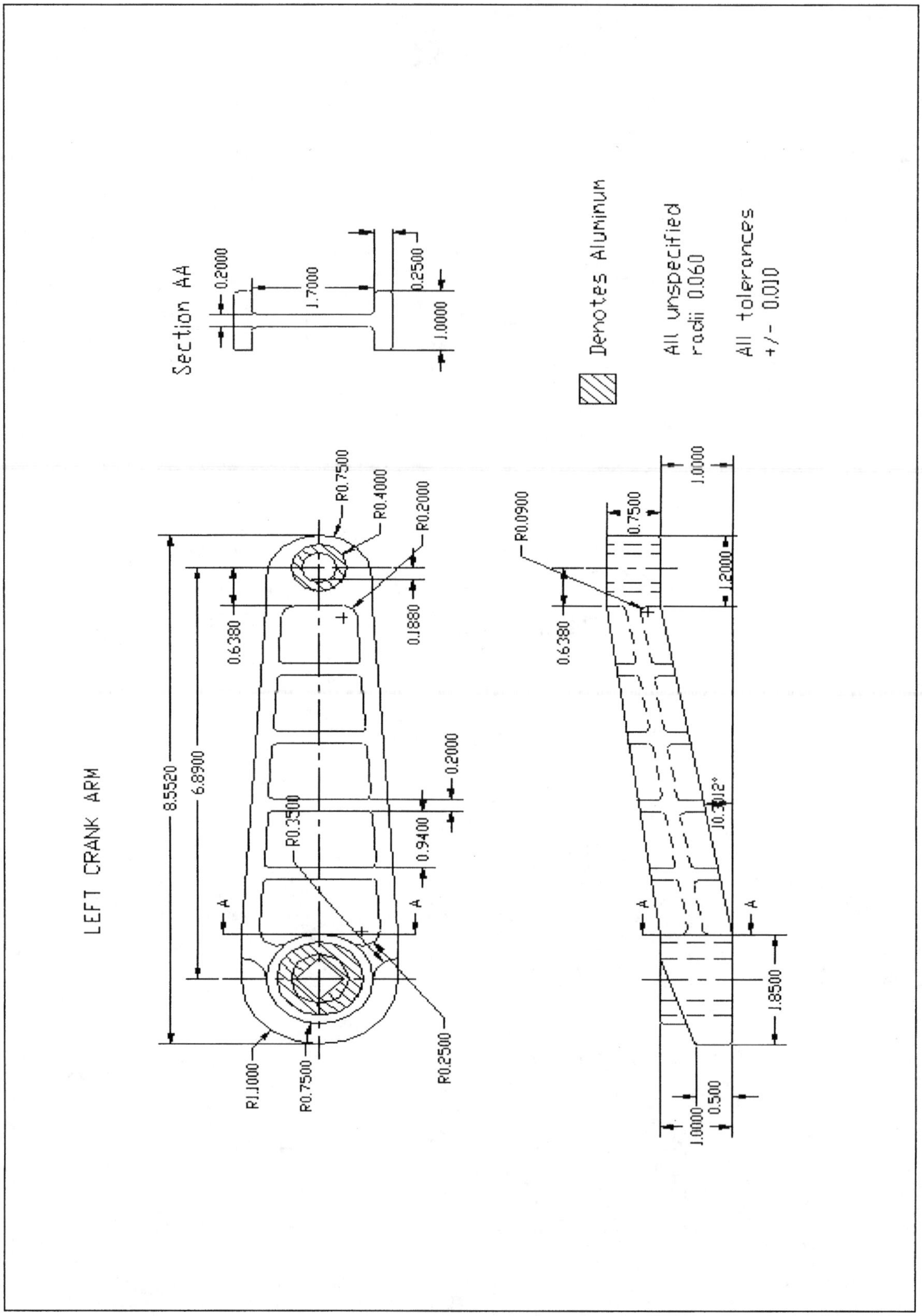

Documented Multiview Drawing (Using CADD Program)

Chapter 13 – The Design Process

The point of learning engineering graphics is not to be able to draw for the sake of drawing or to become a technical draftsman, but to use drawing as a tool in the design process. Ideation drawings are used early in the design process to sort out ideas. Keep in mind that engineering is a visual process, so much of the thinking that is done will be done with sketches rather than with words. Communication drawings are used throughout the design process to share ideas with people from design team members to managers to customers. Engineering design in the modern world is a team activity, so communication and clarity are essential for success. Communication drawings are necessary for a design team to continue to work coherently towards the same goals. Documentation drawings are used to specify all of the details of the design so that it can be properly built as the designers envisioned it. Engineering designers rarely are responsible for manufacturing their design product, so all of the information required to build it must be carefully and thoroughly recorded. All of this, however, does not help to explain what a design process is, or why we are concerned about having a design process at all.

Stated simply, a design process is a way of organizing all of the issues and details that must be considered to allow a design team to work together efficiently, and to produce the best design they can for their given problem. What are some of the issues that a design team must consider while trying to design a new product or process? Take a few minutes to consider the question, and write down everything that comes to mind. It will probably help to pick a specific product or process, such as a new car or a new kitchen appliance. Try to think in terms of general issues, rather than specific features of your product. Table 13.1 is a list of some of the issues that must be considered while trying to design a new product.

Functions	What must the product do?
Materials	What will be used to make the product?
Manufacturing	How will parts be formed?
Assembly	How will the product be put together?
Cost	How much will it cost to make the product?
	How much will the consumer pay for the product?
	How many resources will the product use?
	How will the product affect the environment?
Maintenance	How will the product functions be maintained?
Safety	Is there a chance the product may harm someone?
	Can the product be made so the user cannot be injured by it?
	Can the product be disposed of safely?
Reliability	Can the product be made to work indefinitely?
Operating Conditions	Where will the product be used?
Codes/Standards	What rules exist that must be adhered to?
Impact	How will the product effect the environment and society?
Aesthetics	How will the product look?
Ergonomics	How can the product be designed to fit the human body?

Table 13.1: Some Design Issues

While Table 13.1 is certainly not an exhaustive list, and you may have listed issues not contained here, Table 13.1 does indicate the many diverse and complicated issues that you must consider every time you begin to try to design anything. A design process is an organized structured approach to solving a design problem. By using a design process you should be able to consider all of the issues in Table 13.1, and still come up with the best design solution to your problem in a timely and efficient manner. The remainder of this chapter is the basic description of a design process, as well as an outline of how you will utilize drawing as a tool as you design products and processes.

There are many variations on the design process, but they are all fundamentally similar. The five basic steps of a design process are:

1. State the problem.
2. Develop design specifications.
3. Develop design concepts.
4. Select a design concept.
5. Complete the design details, including documentation and production.

Drawing is an important part of a design process. Ideation drawings are used in the first three steps, as well as during iterations. Communication drawings are used throughout the process, both among the design team and with outside managers and sponsors. Documentation drawings are the core of step five, and no successful or efficient production will occur without complete documentation. Notice also that developing solutions to the design problem does not occur until step three. Steps one and two are for making sure that you have carefully considered *what* you want your design to do before you figure out *how* you are going to do it. This is often hard to convince yourself to do, but think of the effort spent as an investment. If you spend the time to clearly understand the problem, you are much more likely to come up with a good solution, or be able to pick the best solution out of your many ideas. Indeed the entire process begins with being able to state a problem.

Step 1: Stating the Problem

You cannot solve a problem if you do not know what the problem is. This may seem obvious, but your ability to develop a good design hinges entirely on this simple beginning. The problem statement does not have to be long. A sentence or two is usually sufficient to state the basic problem. The problem statement must be clear. It must imply the scope and scale of the problem. Most of all, it must not imply the solution. When you are given the task of solving a problem, you are often told how it should be solved as well. It is important, if you want to be able to consider all possible solutions that you take a figurative step back and ask yourself what *really* is the problem. Let us say, for example, that a customer comes to you and says:

> We need you to design a hoist to lift our shipping boxes onto scales. We need to weigh the boxes to determine shipping charges, but they frequently end up too heavy for our workers to lift safely.

So, what is the problem? Did you say that you need to design a hoist, or was that too obvious? Maybe you said that the problem is that the boxes need to be weighed. If you did, look at the statement again. What is the customer trying to get? Yes, he or she does want the weight of the boxes, but only in order to assess shipping charges. In this case the customer implied two levels of the solution in the statement of the problem. So the actual problem is that you need to assess shipping charges, and the current method is by weighing the boxes. If you state that the problem is determining the weight of the boxes, then you will only look for solutions in that area. If you state that the problem is assessing shipping charges, then you open an entirly new realm of possible solutions, while still giving yourself the option to develop a new method of weighing the boxes. How else might you solve the problem? That likely depends upon what is being shipped. It may be possible to maintain a list of unit weights for various products, and then accurately estimate the weight based upon the products being shipped. The point is not to solve the problem immediately, but to make sure that you understand the problem and give yourself the greatest range of possible solutions. If you state the problem narrowly, you will consider only a narrow range of solutions, and the best ideas might never come up for consideration.

Some people will come to you telling you their problem and how they think you should solve it in the same sentence. Other people will come to you with no idea of what their problem is, but only that they do have a problem. In either case, it is almost a certainty that you will have to gather some information before you can definitively state the problem. There are many different sources of information, but most of your information will come from current or intended customers or users. The most obvious way to get information from people is to ask them. This can be done through surveys to get a small amount of information from many people, focus groups to get reactions to specific issues from fewer people, or interviews to get a larger amount of detailed information from very few people. Along with asking people, however, it is also important to observe them in action. People are not always aware of the small details in their own job tasks. Moreover, humans adapt very well, and usually find methods for working with difficult or poorly design equipment that they come to tacitly accept as part of the procedure without thinking too much about it. It is an interesting exercise to have someone describe the steps in a job task to you and then to either observe them in action to see if they left any steps out, or to try to complete the task yourself from their description. In many cases you will find that they have omitted small details without which the task will be very difficult or incomplete. You will also find in many of these cases that using drawings to explain work procedures instead of relying solely on verbal descriptions can alleviate much confusion.

There are other sources of information aside from people, such as existing designs and devices that you may be trying to improve, replace, or compete against. Before you dive into trying to solve a design problem you should study the current standards in the field, or what is often referred to as the state of the art. One method to learn a great deal about an existing design is to dissect or reverse engineer it, which can also be fun. Another important source of information that is surprisingly easy to overlook is your local library. There are potentially dozens of books, journals, papers, articles, or web-pages available that can quickly provide you with information that would be very difficult or time consuming to produce yourself. You may find that other people have already answered many of your questions or solved very similar problems if you are willing to do a little searching through trade journals and technical papers. So before you rush out to start testing, remember that two weeks in the lab can easily save you two hours in the library!

As a running example, let us say that you have some table tennis playing friends who have trouble finding opponents as often as they would like to. You can phrase this as a problem in many different ways, but for the example let us use the following problem statement: A person cannot play a realistic game of table tennis without an opponent. Furthermore, let us decide to take the approach of designing a table tennis opponent, and at the risk of biasing our approach, let us call this a table tennis robot. We will return to this example in the next section.

Step 2: Developing Design Specifications

Once you have a problem statement, you have begun the process of determining what you want to design. Before you decide how you are going to solve your problem, you need to spend more time developing the details of what your design must do to best solve your problem. This is called developing design specifications. At this point in the design process, specifications are quantitative descriptions of what you want the design to be or to be able to do. In effect, design specifications are boundaries that you will use when generating and selecting ideas in your search for the best solution to your problem. We will develop our specifications by first listing them qualitatively as statements, and then quantifying them. When trying to develop design specifications it is critical that you continue to concentrate on *what* your design needs to do, not how it is going to do it. Once you have developed a set of specifications, you will have a way to compare all of your design concepts against the same standard. Moreover, if you are careful and thorough when you develop your design specifications, it should become clear exactly which design concept most closely meets all of your design requirements. Once again, as engineering design is a visual process, drawings are an invaluable tool for clarifying and communicating your needs and goals.

The first step in developing design specifications is creating a qualitative list of everything that you need or want your design to accomplish. This list should include all of your objectives for the project, and all of the functions that the design must have. You may want to subdivide your list so you can consider each of these topics separately. Objectives are project goals, such as cost, reliability, safety, appearance, or quality. Functions, put simply, are what the device or product needs to do in order to be successful. The primary function of a washing machine is to clean clothes. If a newly designed washing machine cannot do this, then the design is destined to be a colossal failure. Since specifications, objectives, and functions are easy to confuse, they are reiterated in Figure 13.1 for easy reference.

Objectives – Qualitative descriptions of the goals for the design.
Functions – Qualitative descriptions of what the design must do to be successful.
Design Specifications – Quantitative descriptions of the objectives and functions that form both targets for design development and criteria for design selection.

Figure 13.1: Objectives, Functions, and Design Specifications

Let us start by describing our objectives or goals for the functions. Although objectives will eventually be used to form boundaries, you should be aggressive when setting your project goals. The higher the goals you set, the more you are likely to achieve with your design. Once again, objectives are what you want to accomplish with the design. Quality, reliability, safety, aesthetics, ergonomics, and cost are examples of objectives. Most of your objectives may be divided into sub-objectives. For example, one of your objectives might be that the design is "easy to use." "Easy to use" is a very broad statement, so it helps to break this down, defining it more clearly. Three potential sub-objectives for the objective "easy to use" are: few steps to operate, little time to set up, and intuitive to operate. These are not the only possible sub-objectives for the objective "easy to use," and these might not all be relevant to your design problem, so you need to consider what the relevant objectives and sub-objectives are for your specific task. One way to list out all of your objectives is to use a tree. Figure 13.2 shows an example objective tree for our table tennis robot. Notice in this case that ease to use includes that it can be set up without difficulty. This could also be subdivided into several different objectives relating to time required, steps required, and whether or not it is intuitive. In this case, however, the goal is to make sure that one person can set up the robot and run it, since it is intended for use by one person. If this is not true it will not sell. Furthermore, if you do use an objective tree, do not make the mistake of trying to make it symmetric or trying to take each branch to the same level. Not all objectives will have the same number of sub-objectives. You need to keep focused on what you want to accomplish with your specific problem.

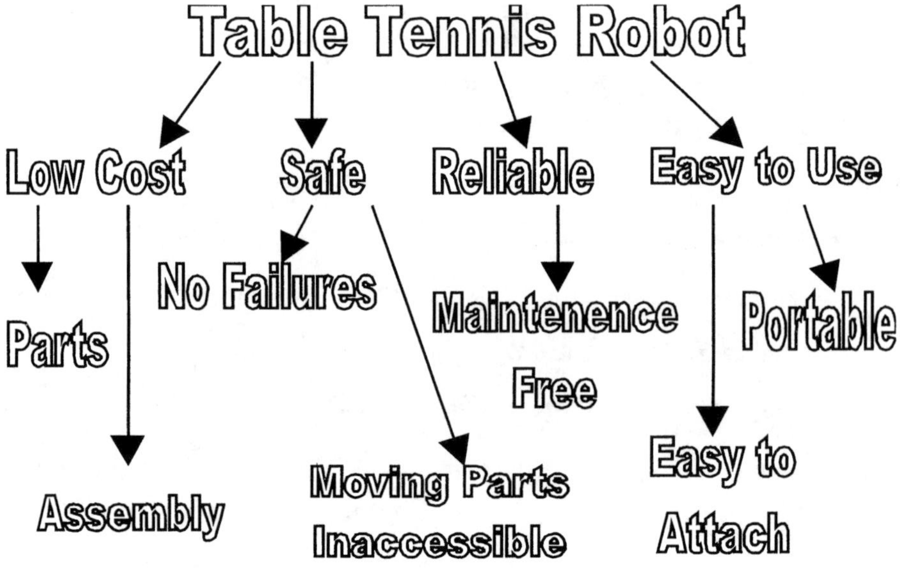

Figure 13.2: Example Objective Tree

Along with listing all of your project objectives or goals, you also need to make a complete list of all of the designs functions. Once again, functions are what the design must do in order to be successful. Some functions are obvious. A camera that does not take pictures will not be a successful design. Most functions, however, are left to the discretion of the designer. Will the camera autofocus? Will it have more than one light metering option? Will it have a built-in flash? Will it automatically advance and rewind film? All of these are potential functions, and as the designer, you must select the functions that you want your design to have. Figure 13.3 shows a sample function list for our table tennis robot. It is important to keep in mind that you are still trying to describe what you want the design to do. It is easy to decide that a function will be too difficult, and therefore you should not list it as a desired function. Having the table tennis robot keep score is an example of something that seems too difficult to do, especially if a deadline is looming in the near future. At this point of the design process, however, you are trying to describe the ideal design for your problem, so exactly how you will accomplish these functions is not yet a matter for concern. Put down the functions that you would like your design to have, even if you have to abandon some later. It is easier to remove functions than to add them once you have begun to generate concepts. Once again, a tree can be used in place of a list if it helps you to visualize and organize more efficiently.

1. Launch Balls
   - Velocity
   - Spin
   - Trajectory
   - Aiming
2. Recycle Balls
   - Catch Balls
   - Temporarily Store
   - Feed Balls to Launcher
   - Release on Time
3. Keep Score
   - Interpret a Hit
   - Display Score

Figure 13.3: Example Function List

Now that you have listed all of your objectives and goals for the functions, you have created a qualitative description of your ideal design.

The next step is to try to quantify each of your goals and functions so that they can be assessed in a simple and unbiased manner. There are two parts to this process. First, you need to determine how important each objective or function is and how you are going to measure it (i.e. what unit you will use). A 1-5 scale usually works well for ranking importance, but you can use a different scale if you prefer. Second, you need to determine what you want your design to achieve, which will give you an ideal performance range, and what you are willing to accept, which will give you a minimal performance or acceptable range. An example for how this might look for some of our objectives and functions for our table tennis robot is shown in Table 13.2.

Specification	Units	Rank	Ideal value	Acceptable value
Cost to Produce	dollars	2	$50.00	$70.00
People for Set Up	number	1	1	1
Time to Set Up	minutes	3	3	5
Velocity Range	meters/second	2	0.5 -- 10	1 – 6
Number of Spins	number	4	5	3
Balls Stored	number	5	24	12

Table 13.2: Sample Specifications

Notice on the second specification in Table 13.2 that the acceptable and ideal values are the same. If you, as the designer, believe that a specification is critical to the design's success, then you need to be clear that the specification cannot be compromised. Safety standards are examples of specifications that usually should not or cannot be compromised. Compromises, however, will be part of the design process. As such, it is important that the ranking given to each specification is truly reflective of the importance of the issue. While it is not true for any of the specifications shown in Table 13.2, sometimes it must be left to human judgement whether or not a specification is being met. Specifying that the design be intuitive is an example. A specification of intuitiveness could be listed as subjective under the units. This may happen on some objectives, and there may be no sensible method for measuring them. In many such cases it is likely that you will have to do some form of testing to determine whether or not you can conclude that your specification is being met. Setting specifications is not necessarily a simple task. You will need to use much of the information that you gathered while trying to determine the problem statement, and probably have to find more information as well. When you are finished, you should have a set of specifications that describe the ideal design for your problem. The specifications will not help you figure out how to design, but they will help you to judge your designs, thus allowing you to pick the one that best meets the needs of your specific problem.

Step 3: Developing Design Concepts

Finally, it is time to start to work on exactly how you are going to solve your design problem. This is the most fun, and also the most challenging part of a design process, and where you really begin to take advantage of your drawing skills. If you are trying to solve a design problem it is likely that there is no obvious solution. This is especially true if it is someone else's problem, as they would have solved it themselves if there were an obvious solution. Thus, you are under pressure to come up with something new and different. You need to be creative.

When it comes to creativity, most of us have crises of confidence. Nonetheless, creativity, liking drawing, is a skill that can be improved upon with practice. Of course, confidence is the first thing that must be addressed. It is very easy to say "I can't" or "No", so easy that many potentially good ideas are squashed at the onset by phrases like "we can't do that" or "that will never work" before there is any consideration of how they might be accomplished. As a simple test, try to name twenty animals or twenty birds. You can do it, even if your first reaction is that you do not know that many. Do a further experiment with your list. Ask friends and family if they can name twenty animals or twenty birds. See how many say no. Show them your list, and see how many of the people who said no are familiar with every entry on your list. The results may surprise you.

Once you believe that you can be creative, you need to begin trying to look at the world in new ways. Problem solving is related to problem stating in that the way a problem is stated can influence your thinking when trying to solve it. Try to solve the two problems below (they are old, so you may have seen them before). Remember that both of these problems require you to break out of implied boundaries to find a solution.

1. Without lifting your pencil off the paper or going through a dot twice, connect the nine dots in Figure 13.4 with four straight lines.
2. The six matchsticks shown in Figure 13.5 are being used to make two equilateral triangles. Without breaking or cutting any of the matches, use the six matches make four equilateral triangles at once so that each side of each triangle is the length of one match.

Figure 13.4

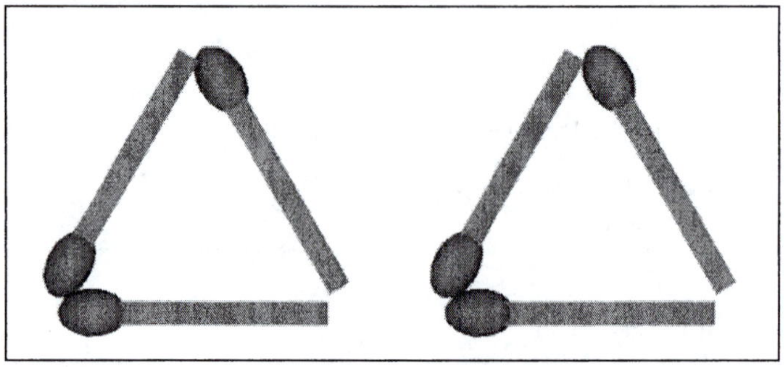

Figure 13.5

The key to solving both of these problems is breaking out from your preconceived notions of what form the solution must take, or "thinking outside the box." In the first problem the dots imply a square that most people try to stay inside. In the second problem the figure implies a two-dimensional problem, so most people try to find a two-dimensional solution. If you are still wondering, by the way, the answers are shown in Figures 13.6 and 13.7. In both cases, the key to solving the problem is stepping back and asking yourself: "What am I assuming about the solution that does not have to be true?" To make this even more insidious, we have all been trained to apply specific methods to specific types of problems, and this can work against our being creative as well. Fortunately, different fields ingrain different biases in their students, so a chemical engineer is likely to take a different approach to solving a problem than a mechanical or electrical engineer. Sometimes you can use this to your advantage, and find helpful ideas in other fields or from people in other fields. This also is another strength of interdisciplinary design teams. Whatever method you use to help you do it, you need to train yourself to think outside the box to be able to come up with creative solutions to design problems.

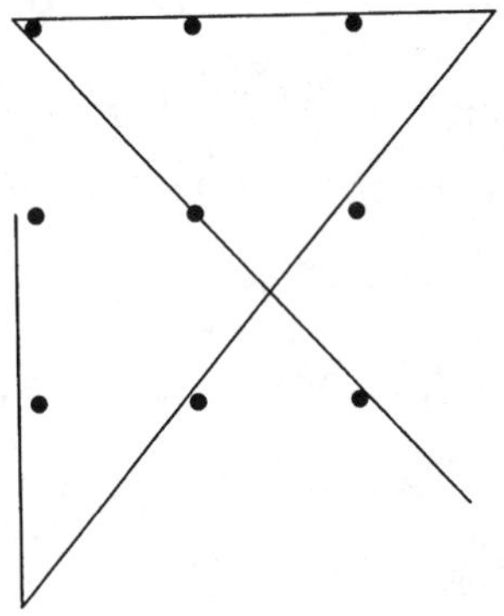

Figure 13.6

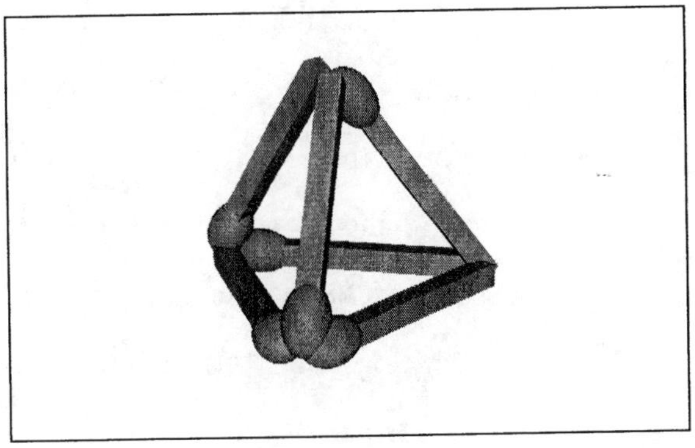

Figure 13.7

As if being creative were not difficult enough, most design problems are also relatively large. So even if you manage to think up a creative idea for one part of your problem, you might be wondering how could you be expected to come up with an entirely creative solution to the whole thing. Trying to design an entire device all at once is a very challenging undertaking. Instead of trying to do it all at once, break it down into smaller pieces. One method is to approach the design function by function. Take each function that the design needs to accomplish, and start listing out ideas for accomplishing just that function. Even better, start drawing possible solutions. Engineering is a visual activity after all, so drawings will help you clarify your ideas both to yourself and to you design team. If you do that for each function, you will end up with a huge chart of ideas with functions listed down the page and potential solutions drawn across it. For some functions you may have ten ideas, while for others you may have only two. As with making an objective tree, symmetry is not important, so sketch every possible solution that you dream up, even if you only end up with one or two for some functions. Once you have drawn all of your different possible solutions for functions, start to mix and match the ideas so that you end up with full concepts. This will let you build full concepts without having to tackle the entire problem at once. The difficulty you may find now is that you can generate too many ideas. If you have four functions and five ideas for each one then you have 625 different combinations! This is an indicator that you need to start to be critical, and only consider combinations that you believe have a good chance of success, otherwise you will run out of time and drive yourself insane trying to select the best concept from a list that is too large. A good way to approach this problem is to use the rankings on your specifications to help you decide which functions to choose first and which ones to add around them. Start with all of your priority one specifications that are based upon functions, and then consider which of your ideas for less important features are likely to work well with your idea for the critical functions. This should allow you to create a reasonable number of complete design concepts.

Step 4: Selecting a Design Concept

Once you have generated design concepts, you need a systematic method for selecting the one that best solves your design problem. There are a number of methods that are not recommended, such as letting the person who yells the loudest have his or her way, or selecting an idea based upon intuition. Instead of these, preferred methods take advantage of the design specifications you spent time carefully developing during Step 2. Since the design specifications represent the ideal design, all you have to do is compare your design ideas to the design specifications in order to find the one that most closely matches the specifications. Of course, real projects are not really this simple, so let us go through three different methods, with increasing complexity and accuracy, for finding the best concept.

The first approach is a Go/No Go analysis. Examine your specifications and select ones that are absolutely critical for the design to be a success, the ones that your design must meet or else it will be a complete failure. Now, make sure that each design concept you are considering can meet each of your critical specifications. If it cannot, then throw it out. This simple step will save you from spending time and energy trying to analyze an idea that is doomed to fail.

Once you have weeded out all of the ideas that were going to be disasters, it is time to start looking more carefully at the ideas that just might work. If you had enough time to really consider potential solutions, you are likely to have quite a few ideas to choose between. What you want to do next is pick to try to pick the best ideas out of the bunch without having to get too in depth with your analysis. One method for doing this is called a benchmark study. Select one design or design concept as a benchmark, and rate every other design concept against it specification by specification, simply rating it as better (+), worse (-), or about the same (0). Your benchmark will usually be a product that already exists, either a competitor's model or one of your own old models. The benchmark study is basically a study to insure that you are making progress in your new design. You cannot expect to have much success if your new design is not as good as what already exists on the market. After rating each design versus the benchmark design, add up the scores of each design, and then select the top two or three designs for further analysis. Also keep an eye out for designs that have a particularly strong area that can be worked into another design, or designs that have a glaring weakness in an otherwise good design that must be solved before the idea has a chance to succeed. This type of analysis is designed to help you get from a large number of design ideas down to the few that really have potential. An example for two imaginary concepts for our table tennis robot is shown in Table 13.3. For the two concepts listed, Concept 1 scores a zero, while Concept 2 scores a +2, indicating that it is a better design based upon the specifications.

Specification	Benchmark	Concept 1	Concept 2
Cost to Produce ($)	0	+	0
People for Set Up (#)	0	0	0
Time to Set Up (Min)	0	-	+
Velocity Range (m/s)	0	-	+
Number of Spins (#)	0	0	+
Balls Stored (#)	0	+	-
Total	0	0	+2

Table 13.3: Example Benchmark Analysis

The problem with a benchmark analysis is that it does not take into account the importance of your specifications. The third level of analysis is a fully weighted scoring. This level of analysis entails some effort. It is relatively easy to determine whether an idea is better or worse than an existing idea when the two can be directly compared. It is not as easy to rank a feature accurately on an absolute scale. In order to accurately rank concepts you will need to do a bit of work to try to predict just how well the design will be capable of performing once it is built. This is why you want to use a benchmark study to cut your pool of ideas down to the top two or three. To do a fully weighted scoring of your design concepts, you need to rank each concept on each specification on an absolute scale, 1-5 or 1-10 for example, and then weight each score by the importance of the specification by multiplying the score by the weighting factor of the specification. Once you total the various scores, the design concept with the highest score should be the one that most closely matches your design specifications, and is therefore the best design solution for your problem. These scores, however, need to be based upon real analysis, so expect to spend a bit of time calculating good estimates for the values. In some cases, analysis on paper or by computer still will not provide sufficient information for an informed decision. In such cases, you may consider building a mock up or physical model of your designs or some portion of your designs. This will allow you to do some testing, let you get feedback from more people, and give you the opportunity to make sure things will fit together as you intend. Having said that, Table 13.4 shows what a fully weighted scoring table might look like for two imaginary concepts for our table tennis robot. Once again, Concept 2 wins out over Concept 1, but now it is clear that it does so because it meets all of the important specifications.

Specification	Weight	Concept 1		Concept 2	
		Score	Weighted Score	Score	Weighted Score
Cost to Produce ($)	4	5	20	3	12
People for Set Up (#)	5	4	20	4	20
Time to Set Up (Min)	3	1	3	4	12
Velocity Range (m/s)	4	2	8	4	16
Number of Spins (#)	2	3	6	5	10
Balls Stored (#)	1	5	5	1	1
Total			62		71

Table 13.4: Sample Weighted Scoring Table

As always, it is never too late, even at the stage of concept selection, to reevaluate a design or part of a design and try to improve it. The design process is iterative. You will not be able to go through every step in order without returning to revise and reconsider some of your decisions in prior steps. Therefore, it is always good to look back at what you have done and make sure that you are still satisfied with the path that you have chosen. Until you have committed yourself to production, there is always some room to alter a design. Keep in mind, however, that it is always easier to change a design while it is still on paper. Once you begin to build even prototypes, fundamental changes become harder and harder to implement.

Step 5: Completing the Design Details

The old expression that the devil is in the details rings very true at this point of the design process. Now that you have selected the concept that you have determined will best solve your design problem, it is time to actually build and test it. This requires going through and figuring out exactly what processes and procedures will be used in the manufacturing and assembly of your design. Hopefully, you have been considering issues such as manufacturing and assembly throughout the design process, so you should not be starting from scratch on any of these issues. Still, the final development of a product from a conceptual design usually involves solving some unforeseen problems along the way. A standard path from concept to product usually involves a number of physical models and prototypes.

With the advent of modern solid modeling packages, computer models, which will save both time and money, can replace some physical models, but in most cases at least one prototype will be available. Like every other part of the design process, you should carefully plan out your prototypes before leaping into the workshop and starting to build. Prototyping is expensive and time consuming, so in this sense it should be seen as an investment in the future success of the design product. Before you build a prototype you should have a schedule for its completion and testing, a testing plan that details exactly what you expect to learn about, and a plan for implementing changes to the design based on your testing results before you move on the next prototype or build. If you do not have a chance to make changes based upon what you have learned from prototype testing due to time constraints, you probably did not spend your time and money wisely. There are many different reasons to build a prototype. You can test product functions, product ergonomics and aesthetics, product quality, and the processes you will utilize in the manufacturing of the product. In most cases you will make more than one prototype, and if you plan intelligently, you will be able to learn a great deal and improve your final product through prototype testing.

Throughout the prototype testing process you will move closer to your final design. At all levels of the design detail stage you will need to keep careful and thorough documentation drawings of your design. You will need to document each and every part carefully. The documentation will include the basic shape and size of the part. It must also include the material that will be used to make the part, as well as the forming and finishing methods to be used, and the tolerances to which the part must made. You will also need to create assembly drawings that show how the product is to be assembled. In most cases you will not be part of the manufacturing team that will build the design, so all of the details necessary for its construction must be part of your design documentation. If they are not you will find that the design documentation will return to your desk with requests for more or clearer information, or someone else who may not understand your intent will try to make the necessary changes themselves. As neither of these options is desirable, it is your responsibility to make sure that all of the design documentation is completed correctly before you send it off to the manufacturing floor.

References

Ertas, Atila and Jesse C. Jones, *The Engineering Design Process*, John Wiley & Sons, New York, 1993

Ferguson, Eugene S., *Engineering and the Mind's Eye*, MIT Press, Cambridge, 1992

Parnes, Sidney J., *Visionizing*, The Creative Education Foundation Press, Buffalo, 1992

Ullman, David G., *The Mechanical Design Process*, McGraw-Hill, New York, 1992

Ulrich, Karl T. and Steven D. Eppinger, *Product Design and Development*, McGraw-Hill, New York, 1995

Chapter 14 - Example Design Problem(s)

Situation Analysis

Computer Work Area: The area in question consists of a monitor, key board, mouse, speakers, and microphone on a computer desk.. The CPU sits on the floor under the desk and the keyboard is contained in a sliding keyboard rack. The monitor and speakers sit on a monitor stand which in turn sits on the top of the desk. A separate stand holds the printer and extra printer paper and a telephone. There are numerous items on the desktop in addition to the monitor stand including papers, folders, note pads, assorted writing implements, disk storage units, computer manuals and texts, a telephone directory, CD's, and assorted office supplies. The desk top is visually unappealing, unorganized, and some items are difficult to find.

Problem Statement

Create a way of organizing the items associated with the computer work area in a visually pleasing manner such that the computer operator has easy and efficient access to needed information and materials.

Design Objectives

- Aesthetically Pleasing
- Materials and/or Finish that Coordinate with Existing Desk
- Integrated Design
- Easy to Prototype

Design Function Specifications

- Papers and Folders Organized and Easily Accessible
- Books and Manuals Stored and Accessible
- CD's Stored and Accessible
- Storage for Computer Disks
    - Program Disks
    - Data Files
    - Frequently Used Disks (Current Projects)
    - Frequently Transported Disks
- Organization of Office Supplies

Information Gathering:

A number of office supply and office furniture stores were visited in order to determine the feasibility of using commercial available products to solve the problem or at least to gain ideas for solving the problem. During this investigation it was determined that a product or design solution which totally integrated all of the needs specified was infeasible. However, products were available which offered a partial solution. These products were purchased, put into use, and the situation reassessed. The storage and organization of papers, and the organization of office supplies portion of the problem were satisfactorily met. However, the storage of computer disks was only partially met and the storage of computer manuals and books portion of the problem was left unsolved. It was decided that computer discs could be stored in existing containers except for those which were in constant use. Based on the degree of progress attained at this point, the problem was restated as follows:

> Create a way of organizing computer manuals and help books and CDs associated with a computer work area in a visually pleasing manner such that the computer operator has easy and efficient access these materials.

Revised Design Objectives

- Aesthetically Pleasing
- Materials and/or Finish that Coordinate with Existing Desk
- Easy to Prototype
- Reasonable Cost

Revised Design Specifications

- Books and Manuals Stored and Accessible
- CD's Stored and Accessible
- Integrated Design

Revised Information Gathering

Additional information gathered included:

Size of a CD Case = 4.875 X 5.5

Maximum Size of Computer Manuals and Texts = 8X10

Approximate number of CD's needing storage = Dictated by space required for books

Approximate number of floppies needing storage = 100

Approximate number of books needing storage = 16 inches worth

Idea Generation:

The design problem continues to evolve even at the very beginning of the ideation phase and continues to do so throughout the entire process.

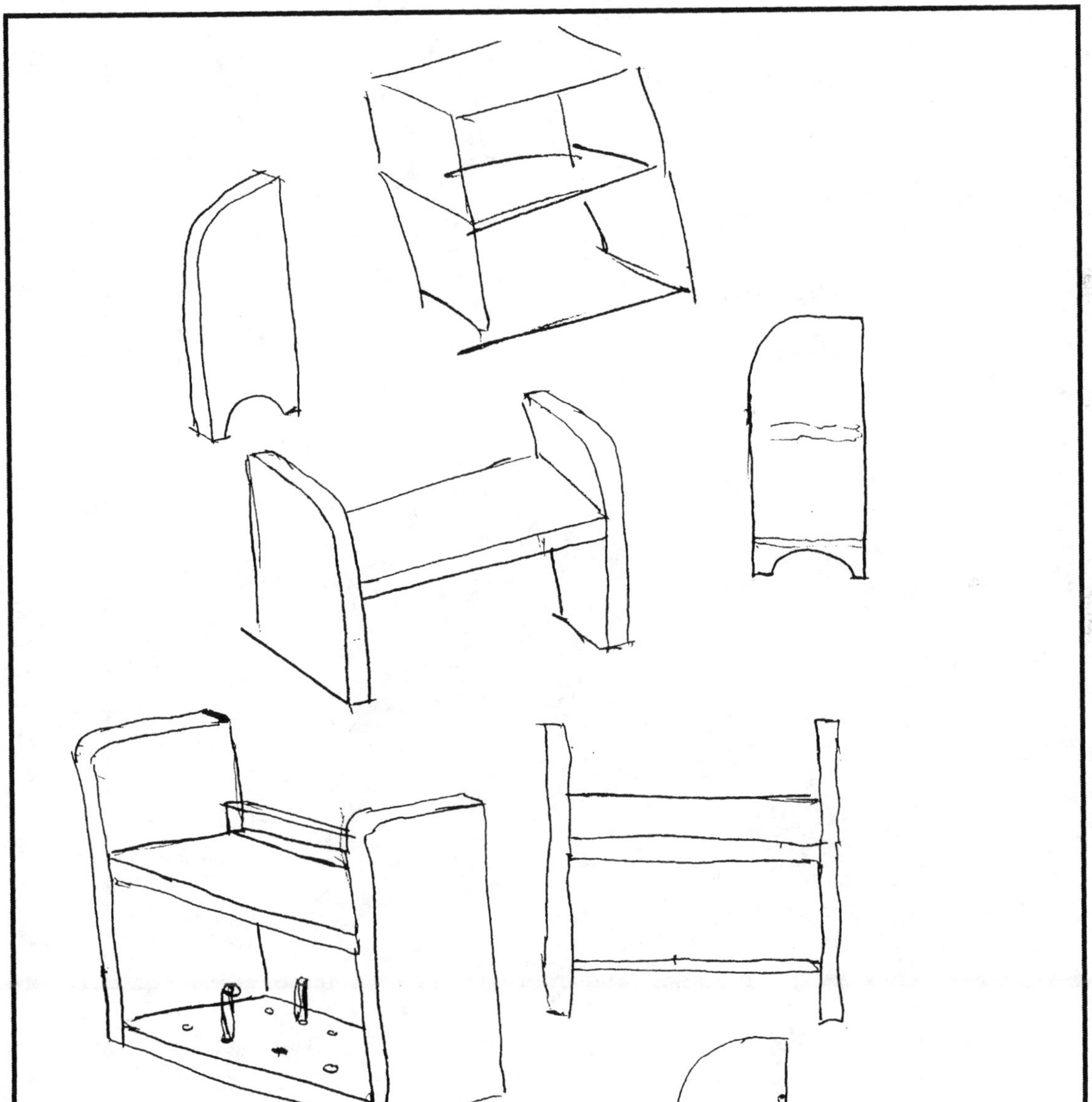

First Attempt at Idea Generation

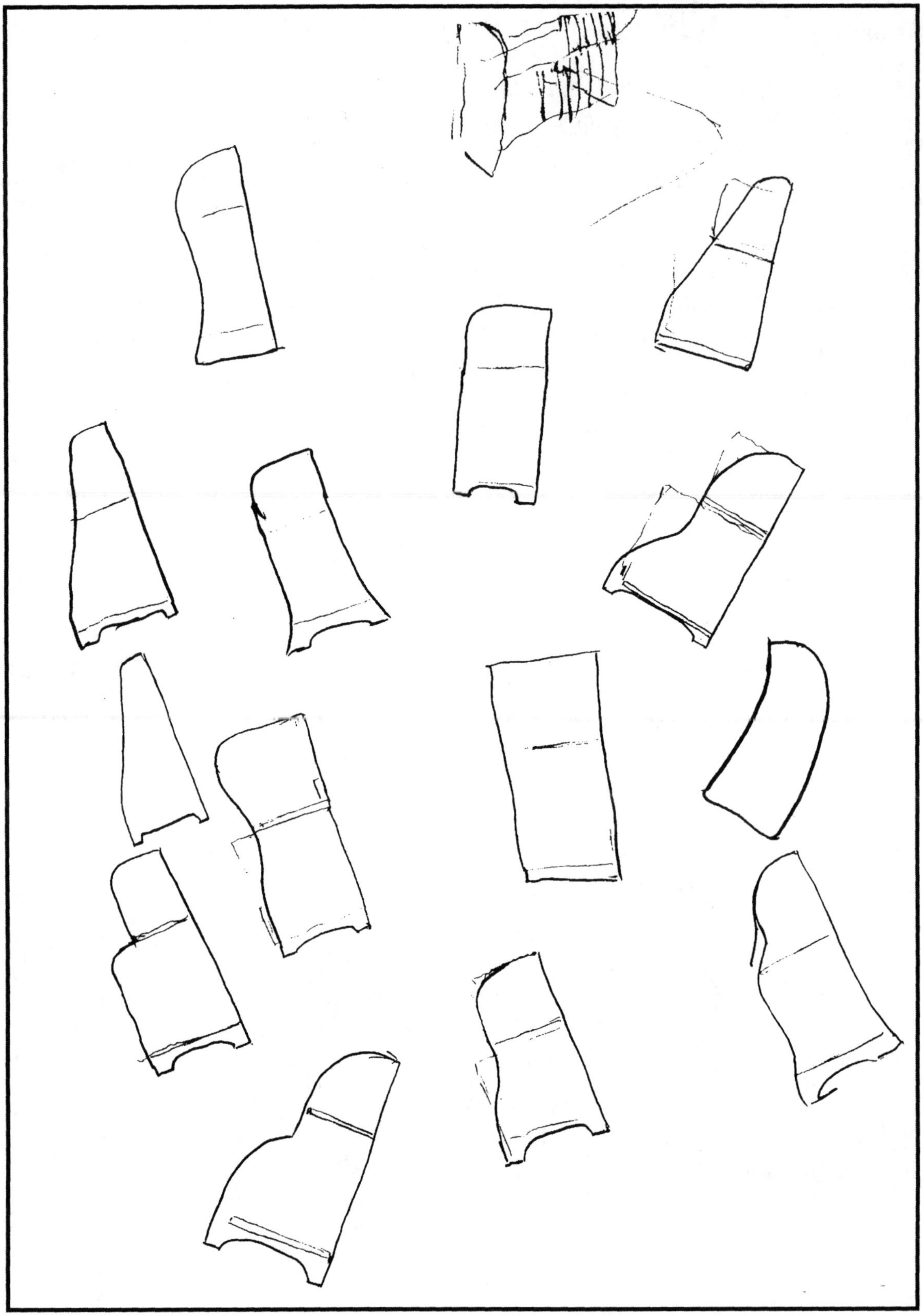

More Ideas

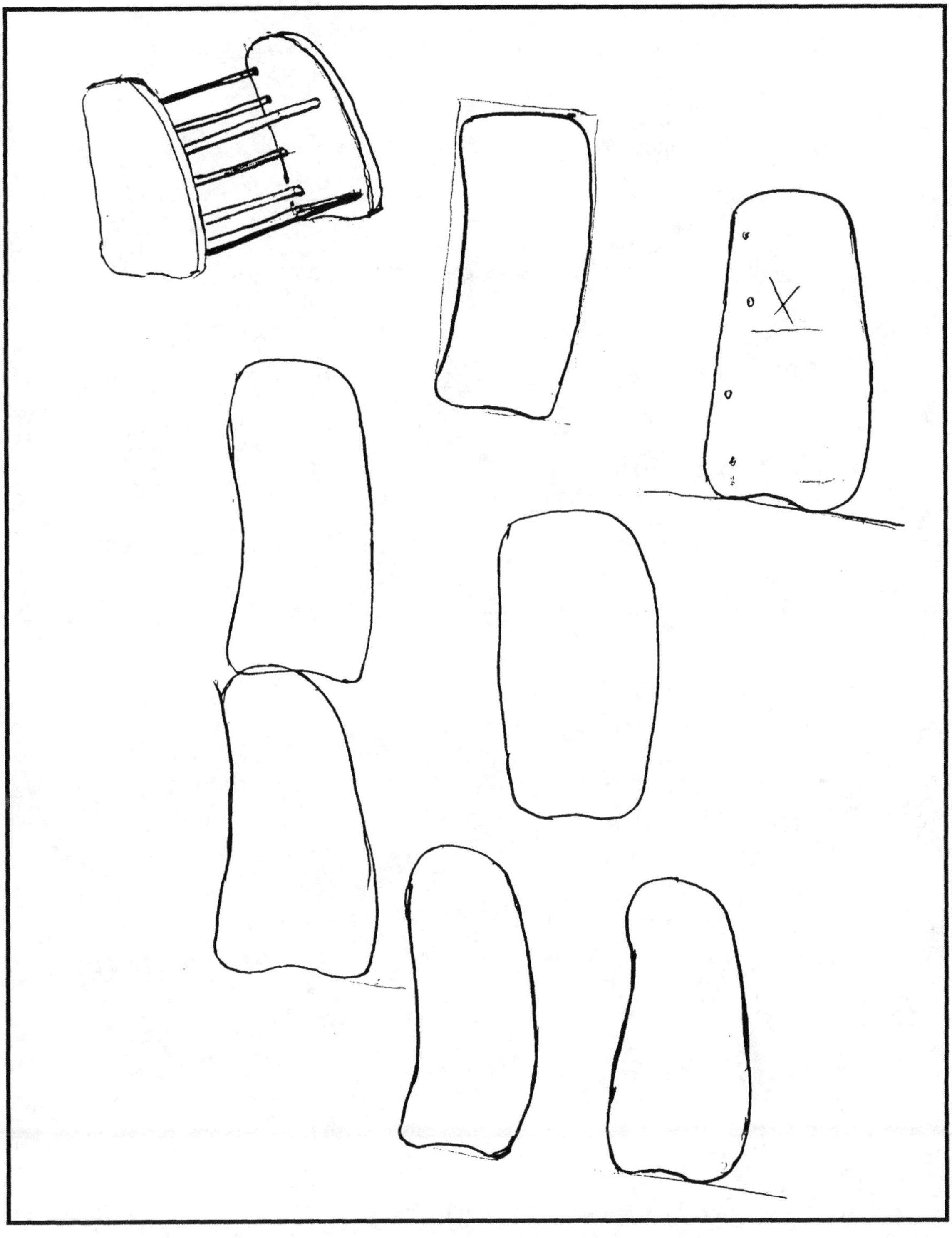

Final Idea Generation

This illustration represents the final phase of idea generation for this problem. The three pages of ideas represent a progression of ideation. This third page represents attempts to find a shape for the side of the device represented here. The shape with the "X" is the one chosen for the final design.

Using Iteration for Communication Drawings:

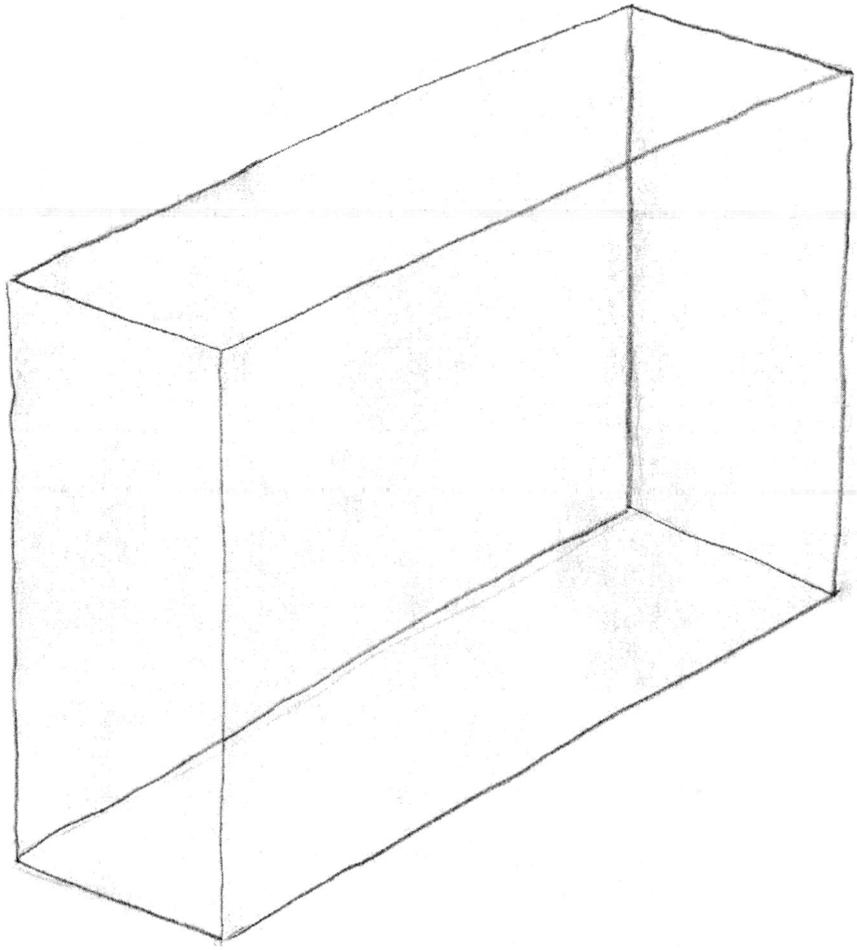

First Iteration

Since the design is continuing to evolve, proportions and size do not have to be exact. At this point in the process, it is important to communicate the design intent so that it can be communicated with others. The details will emerge in the documentation drawings and in the prototype phase of the design.

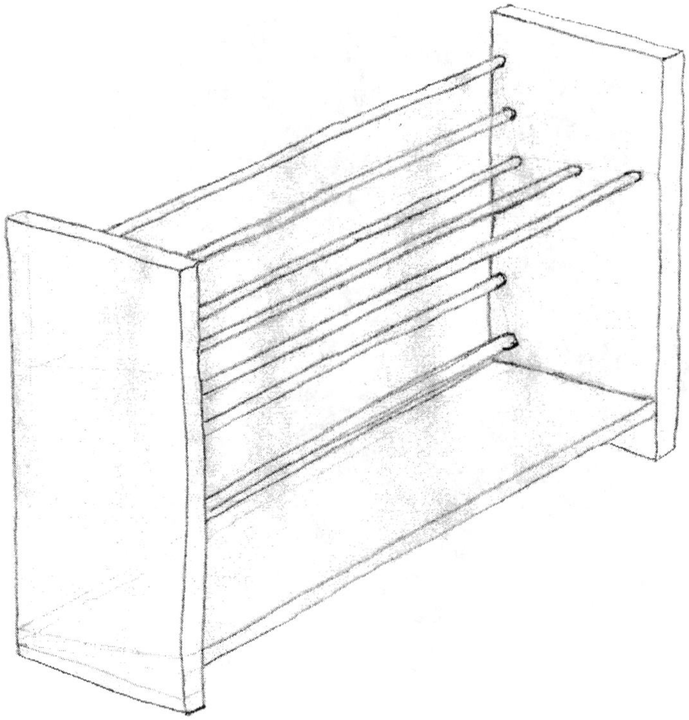

Second Iteration

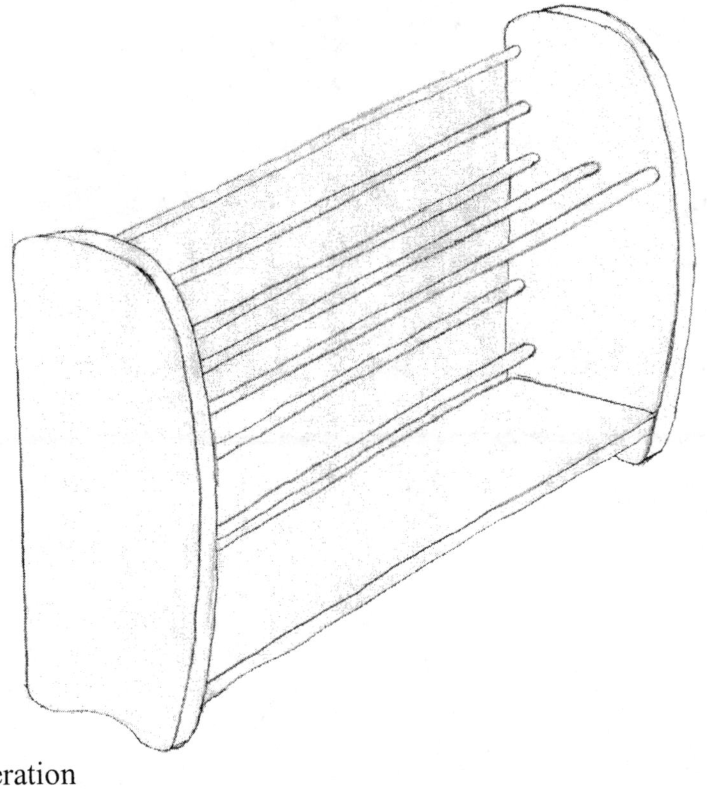

Third Iteration

Computer Generated 3-D Model

The computer and appropriate software can be an invaluable aid to working out details of size, proportion, and shape. A 3-D model can be oriented in any position and rendered in a variety of ways in different lighting situations so that a full understanding of the design may be gained at this time. It should be recognized that the design never really becomes final. A good and conscientious designer can always evolve the design further. However, the real intent of design is not just to explore ideas and "push the envelope" to new levels of creativity. Somehow, all of this eventually has to pay off. Unless we are self employed and independently wealthy, we need to get the design to market, so there has to be a limit to how far we go. At some point, we must quit designing and begin producing. Before we can do that, a prototype should be built to test design function and evaluate the aesthetics properties of the design.

The type of product will determine this. Prior to prototyping, final details for each part and how they are to be assembled must be worked out. After these and production details have been determined, and a decision to move to production has been made, a final documentation drawing will be created.

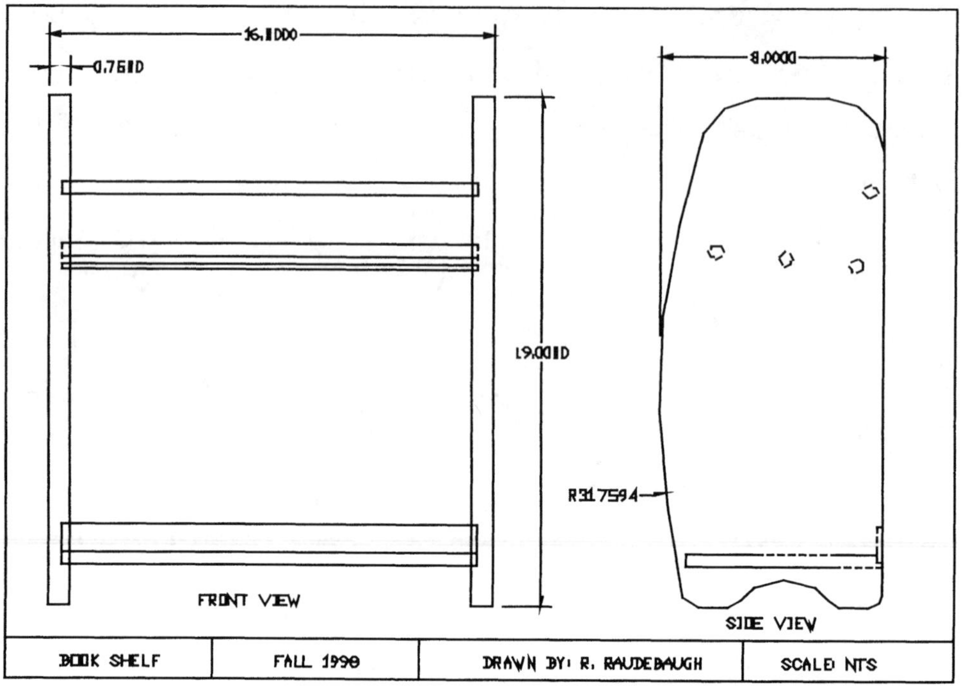

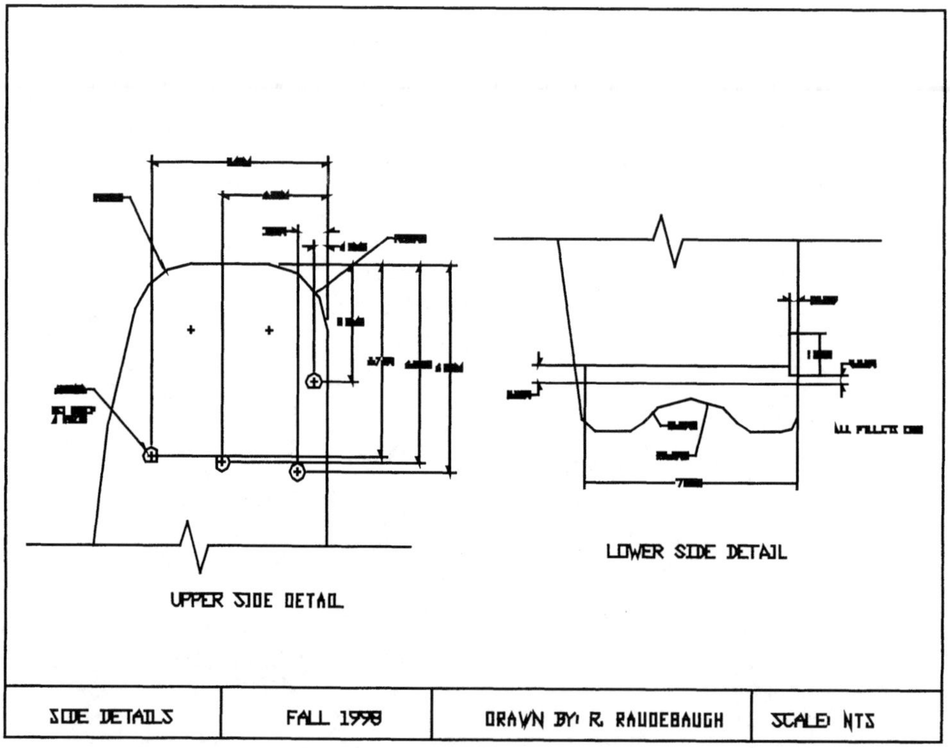

Prototype Drawings of Bookshelf

Appendix A

Example of Student Work (By Permission)

Home Entertainment Center

Project by Anneliese Seyler

ETEC 110 - Engineering Graphics I
Section #27004 - Final Exam
Professor Raudebaugh
December 4, 1998

Purpose of Project:
The following project was designed to be space-saving, functional, and aesthetically pleasing. The entertainment center needed ample storage space, as well as specific areas to hold a 60" television, a VCR or DVD player, a stereo, an amplifier, and at least two stereo components. It needed to be big enough to hold these parts, but small enough to fit nicely on one wall in my home. Based on the above parameters, I designed the following home entertainment center using sketching, iterative development, technical drawing, a computer modeling program (Rhino), and a computer aided drafting program (Intellicad).

Step 1: Design Idea

Exploring Ideas, Initial Sketching

In this initial phase, brainstorming was essential. I did not have an idea of overall shape or dimensions of the product, I did not know how I wanted to orient the components relative to each other. This phase was where design creation began, by experimenting with various and multiple ideas for what my final product would become. After sketching many ideas, I began to get a better feel for what I wanted the final product to took like. I chose the idea that best suited the parameters outlined in the purpose of this project.

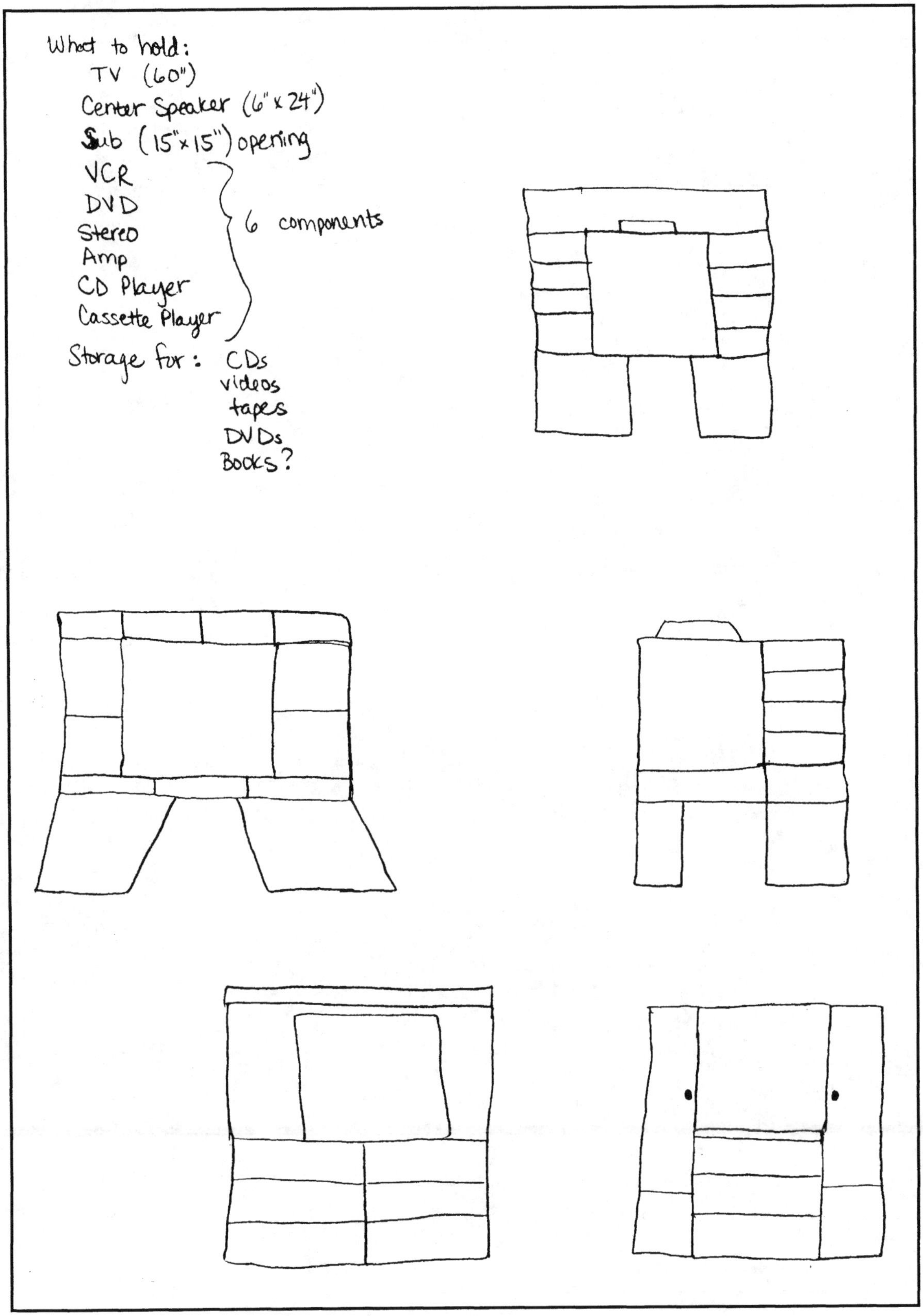

Idea Sketches

More Idea Sketches

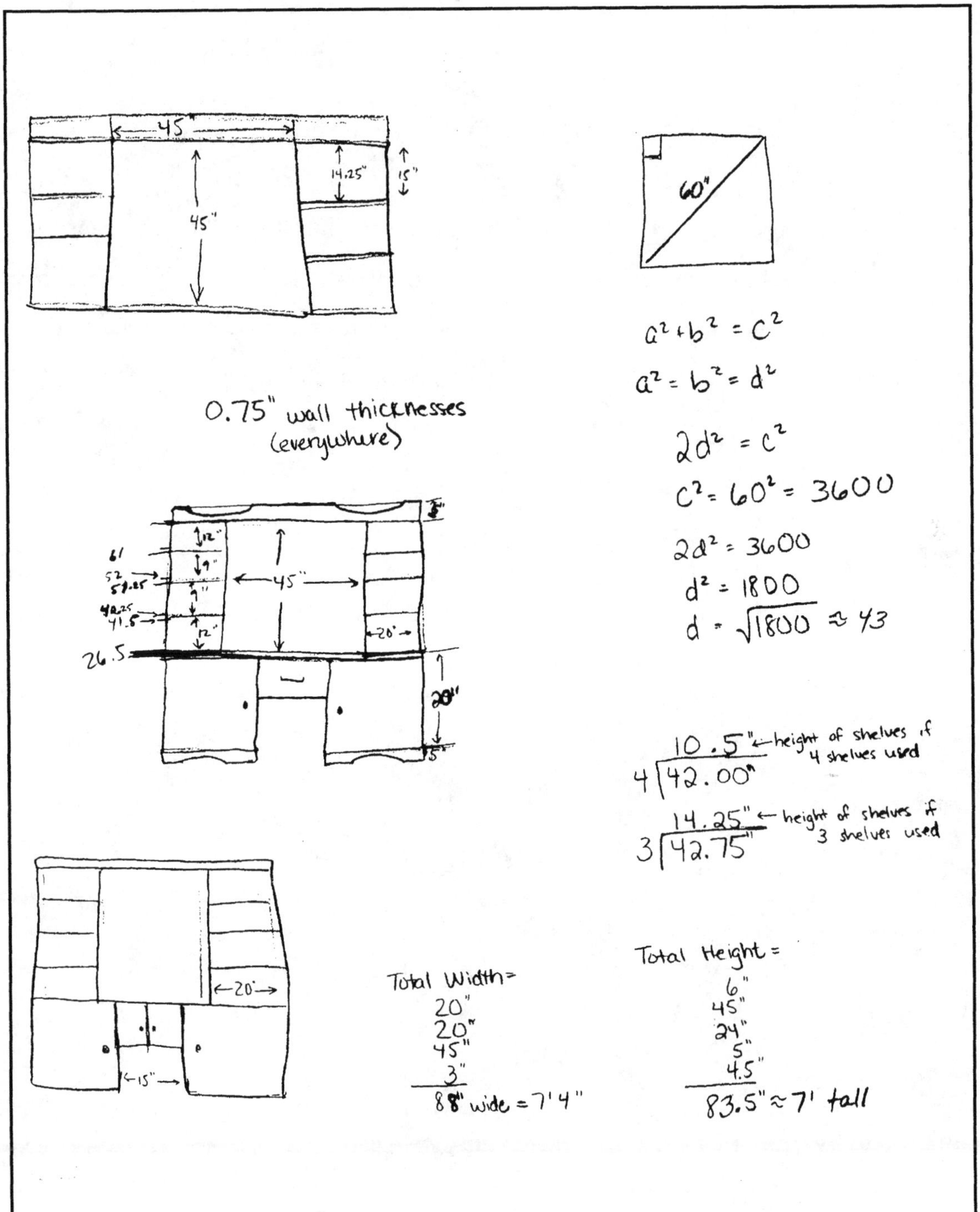

Development Sketches

Step 2: Design Development

Iterative Development of Idea

After choosing which design idea to use, I could develop the design. I now had an idea of what the final part would look like, but details needed to be worked out. Using the iteration process, I used the overall proportions of the product to outline the general shape. Then, the shape was refined through two further iterations. The final drawing in the iteration process includes details which are enhanced by shading. Once the design had been fully developed through this stage, the dimensioning and refinement began.

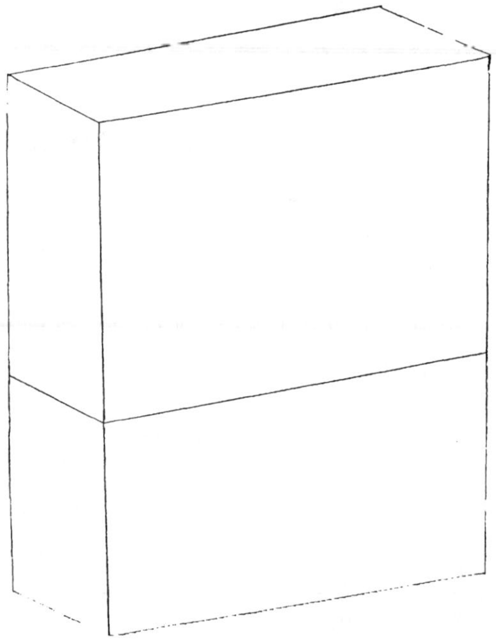

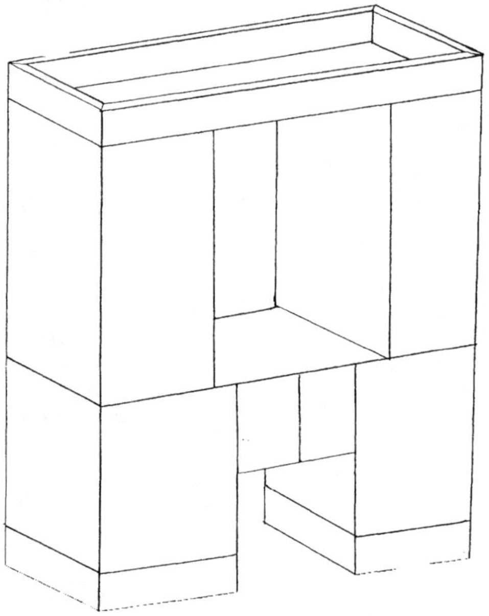

Iterations One and Two

Final Iteration

Step 3: Design Refinement

Technical Drawing, Computer Modeling
Final Dimensioning

In the refinement process, a freehand multiview assembly drawing (technical drawing) was made, from which computer drawings were generated. The technical drawing fully dimensioned the product. Then Rhino, a 3-D NURBS Modeling program, was used to model the product. I generated a 3-D model both with and without the doors, to show the inside shelving in the unit. Next the model was converted to 2-D and imported into Intelicad. Using Intelicad, a multiview assembly drawing was made which shows basic overall and location dimensions. With this, the design process was complete, and the final product is ready to be produced.

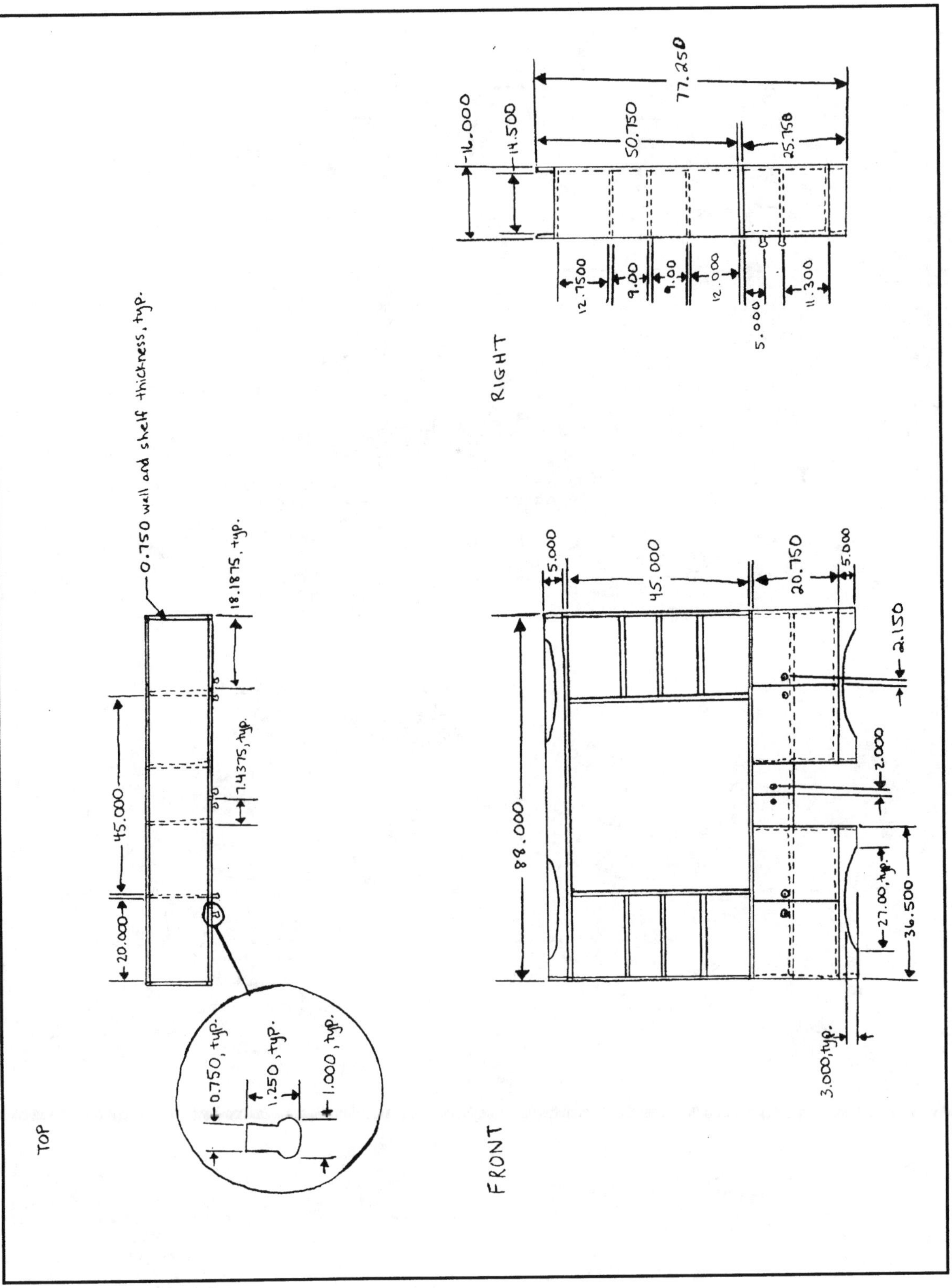

Technical Sketch

Wire Frame Model
(using Rhino)

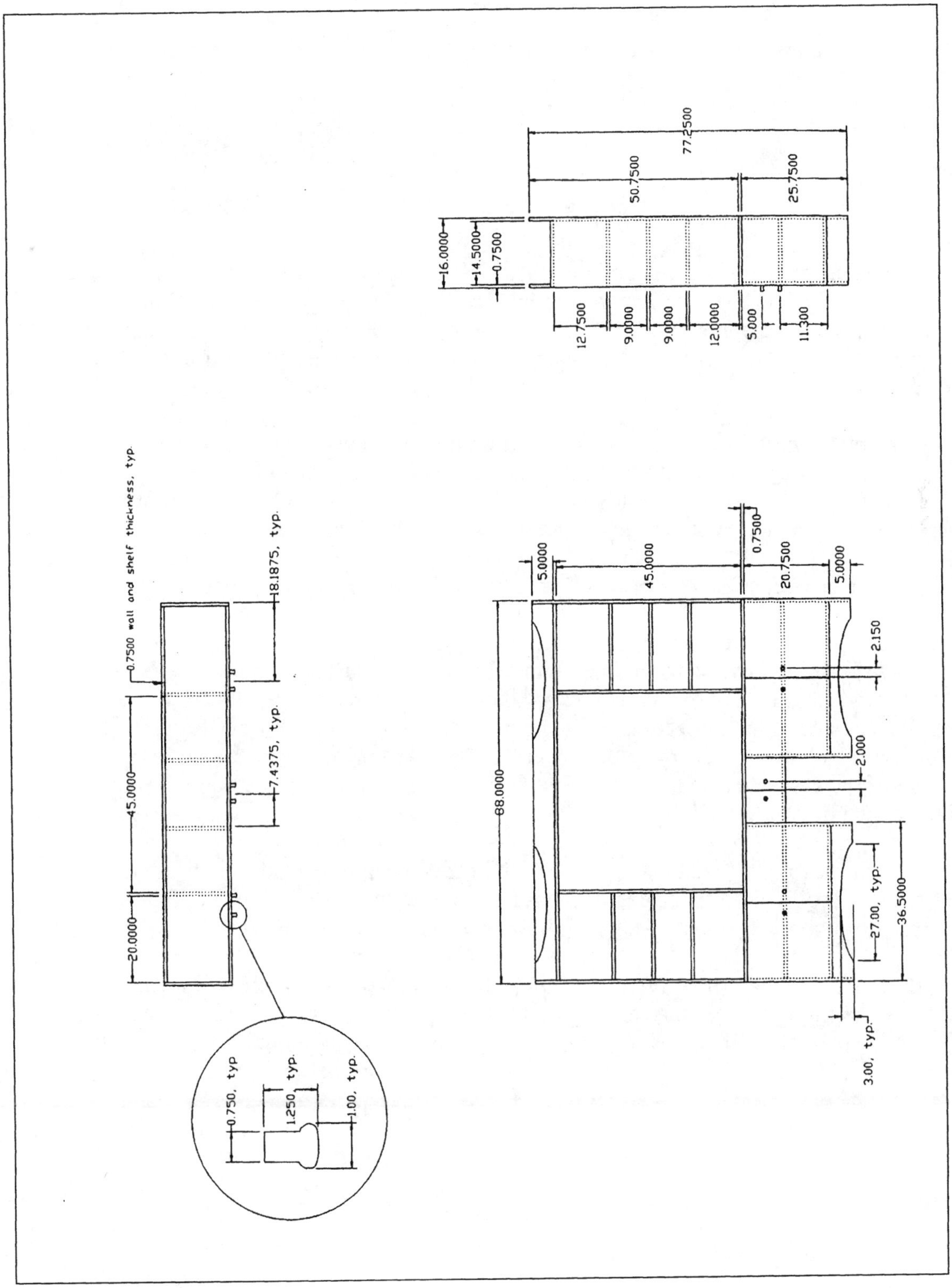

CADD Documentation Drawing

Appendix B

Example of Student Work (By Permission)

Ryan Jinks
ETEC 110
Flashlight

I have designed a functional and inexpensive flashlight that uses standard D size batteries as its power source. it is designed to be ergonomic in that it is shaped to fit the human hand and has a magnet built in to assist the user in holding the light. my flashlight is also designed to be pleasing to the eye, using a streamlined shape and multi-colored plastic construction.

A magnet is cast into plastic and attached with two screws to the side of the flashlight. This is intended to protect surfaces such as painted automobile fenders when sing -the magnet for hands-free operation of the flashlight. The case of the flashlight is contoured to fit the hand, and the switch is readily accessible to the thumb.

The power circuit consists of two D size batteries which are inexpensive and readily available. A standard bulb is also used. Copper strips, a spring and a metal bulb retainer make up the rest of the circuit. A plastic knob connected to one of the copper strips allows the user to switch the light on and off. A strip is fastened to the case by a screw and comes into contact with the bulb retainer, which is in contact with the outside of the bulb. The end of the bulb contacts the positive terminal of the upper battery, which is held in place by spring pressure on the lower battery. The spring snaps into a groove in the bottom of the case and also he p to retain the final copper strip, which completes the circuit by contacting the switching strip.

A threaded ring is used to hold the lens and reflector into the end of the case, and a metal retainer threaded on to the end of the reflector to hold the bulb in place. Batteries are inserted through the open end so that another end cap is unnecessary.

My flashlight is made of common and inexpensive materials, which will keep costs down. Also, the multi-colored plastic case will be rugged and adds a pleasing appearance to the light. All of these features add up to an inexpensive, ergonomic, and attractive flashlight.

Ryan links

ETEC 110

Flashlight

Problem

Statement

Design a flashlight that is ergonomic, functional, inexpensive, attractive, portable, and uses a standard power source.

Design specifications

1. Ergonomic Design - does it fit the human hand?

2. Inexpensive - Are components readily available or easily manufactured?

3. Rugged construction - Is it made from materials that will take some abuse?

4. Standard Power source - Does it use readily available and inexpensive batteries?

5. Functional Design - Does it have features which assist in using the flashlight?

6. Simple Design - Does it use few components?

7. Portable - Is it small enough to carry?

8. Attractive Style - Is it pleasing to the eye?

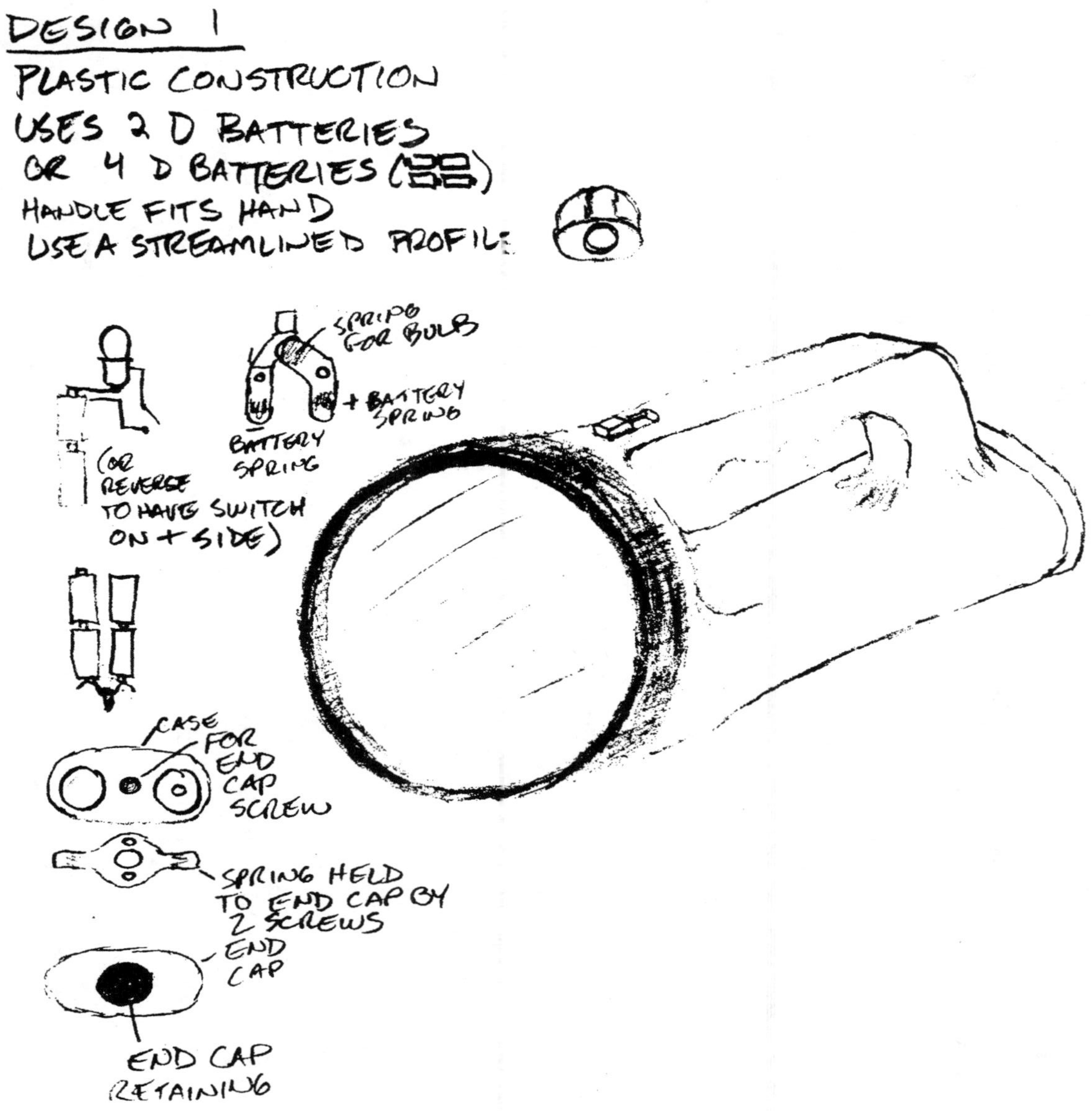
DESIGN 1
PLASTIC CONSTRUCTION
USES 2 D BATTERIES
OR 4 D BATTERIES
HANDLE FITS HAND
USE A STREAMLINED PROFILE
SPRING FOR BULB
+ BATTERY SPRING
BATTERY SPRING
(OR REVERSE TO HAVE SWITCH ON + SIDE)
CASE FOR END CAP SCREW
SPRING HELD TO END CAP BY 2 SCREWS
END CAP
END CAP RETAINING

DESIGN 6

PLASTIC CONSTRUCTION

MOLDED TO FIT A HAND (RIGHT HANDED)

PUSH BUTTON SWITCH

USES 2 D BATTERIES

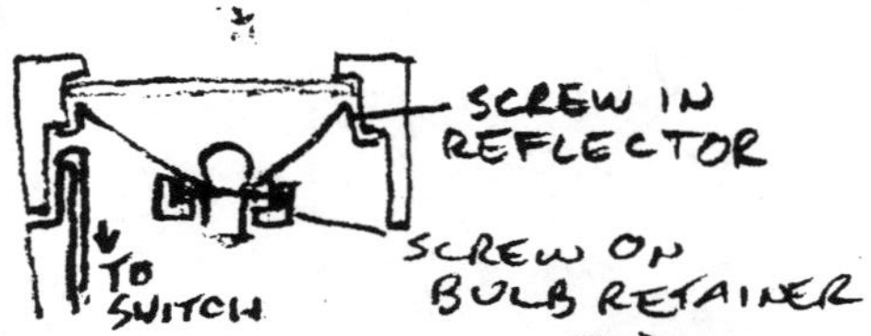

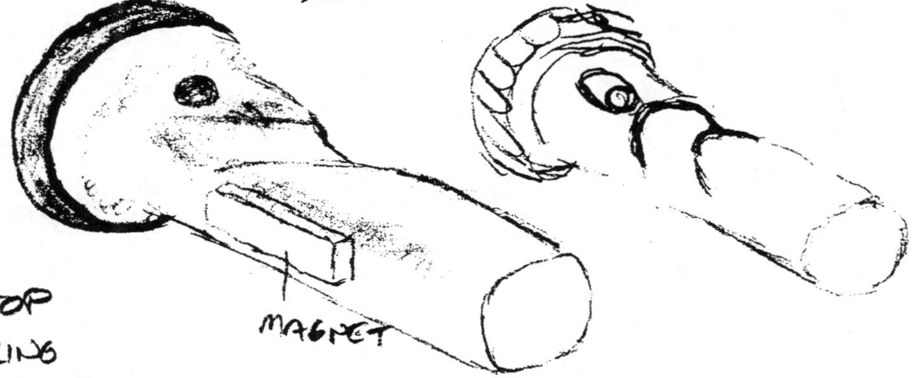

METAL REFLECTOR GROUNDS BULB TO A METAL STRIP WHEN TOP IS SCREWED ON, + SPRING PRESSURE PUSHES + TERMINAL INTO CONTACT W/BULB.

MAGNET ON SIDE OF FLASHLIGHT? (STRONG ONE INSIDE PLASTIC SO IT DOESN'T SCRATCH SURFACES?) YOU COULD GLUE IT INTO CASE

DESIGN 3

PLASTIC CONSTRUCTION

USES 2 D BATTERIES

ANGLED HEAD IS MORE NATURAL POSITION FOR HAND & WRIST TO SHINE LIGHT FORWARD

RIBS ON END CAPS GIVE GRIP WHEN REMOVING

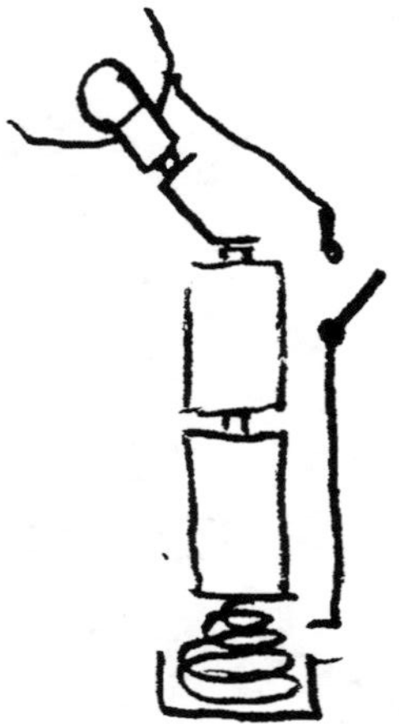

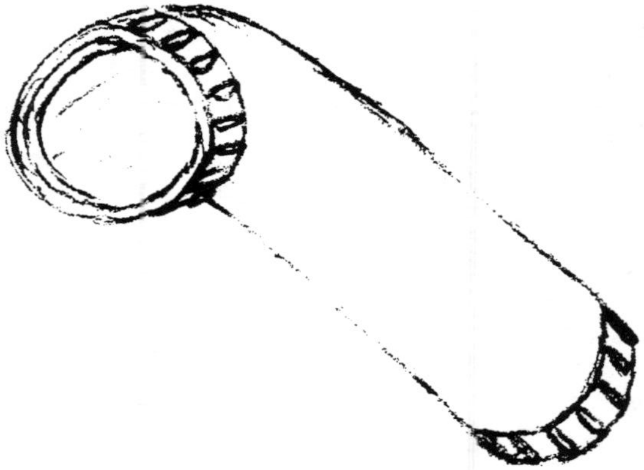

Ryan Links
ETEC 110
Flashlight

Design Selection

I decided to use design idea two for my final flashlight design. I based my decision on caref my specifications. I analyzed how well each design idea met each specification, and then ranked th specifications where my designs differed. By assigning a score to each design idea and weighting tl importance, I decided my second idea was the best of the three, although all of my ideas had merits specifications.

Ryan links
ETEC 110
Flashlight

Design specifications -Compare Design Ideas

Specification	Metric	Design I	Design 2	Design3
1. Ergonomic	1-5 (I min)	4	4	3
2. Inexpensive	1-5 (I min)	3	4	5
3. Rugged	yes/no	yes	yes	yes
4. Standard power	yes/no	yes	yes	yes
5. Functional	# of features	1	2	1
6. Simple	# of parts (estimated)	17	10	11
7. Portable	is it under 12" long?	yes	yes	yes
8. Attractive	1-5 (I min)	5	5	4

Specification	importance (1-5, I min)	Design I Score	Design I weight	Design 2 Score	Design 2 weight	Design 3 Score	Design 3 weight
1. Ergonomic	5	4	20	4	20	3	15
2. Inexpensive	3	3	9	4	12	5	15
5. Functional	5	4	20	5	25	4	20
6. simple	3	3	9	5	15	5	15
8. Attractive	4	5	20	5	20	4	16
Total			78		92		81

FLASHLIGHT DESIGN

FEATURES:
- → ERGONOMIC SHAPE
- → MAGNET

CONSTRUCTION
- → PLASTIC

POWER SOURCE
- → 2 D SIZE CELLS

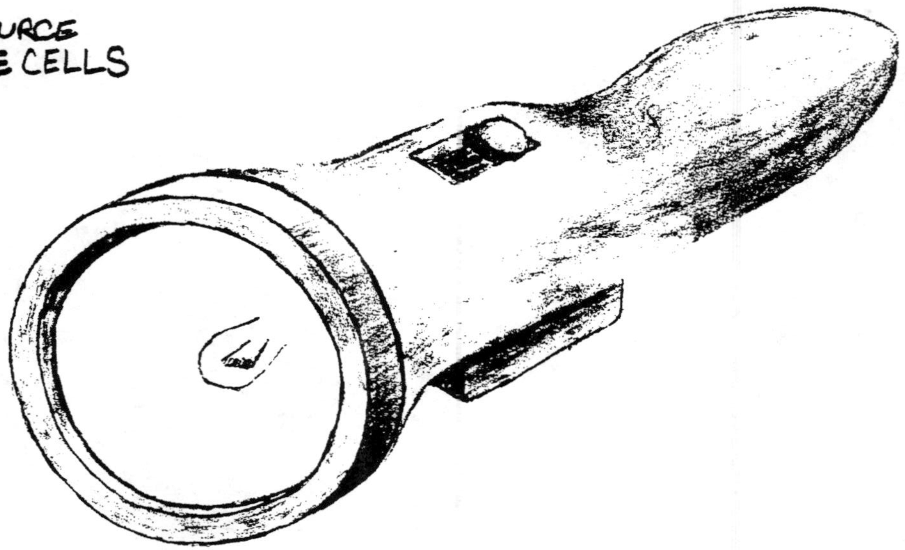

RYAN JINKS
ETEC 110

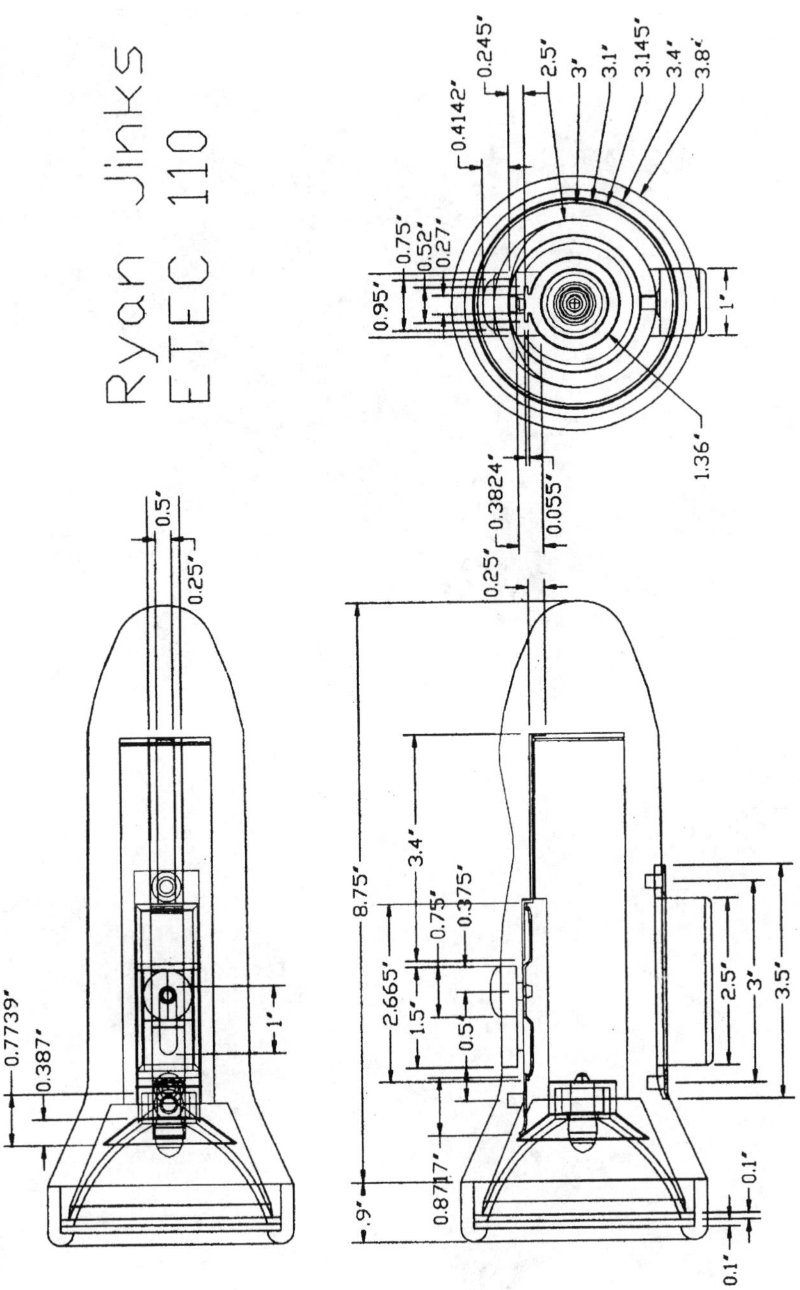

Documentation Drawings done in CADD

3-D Computer Models

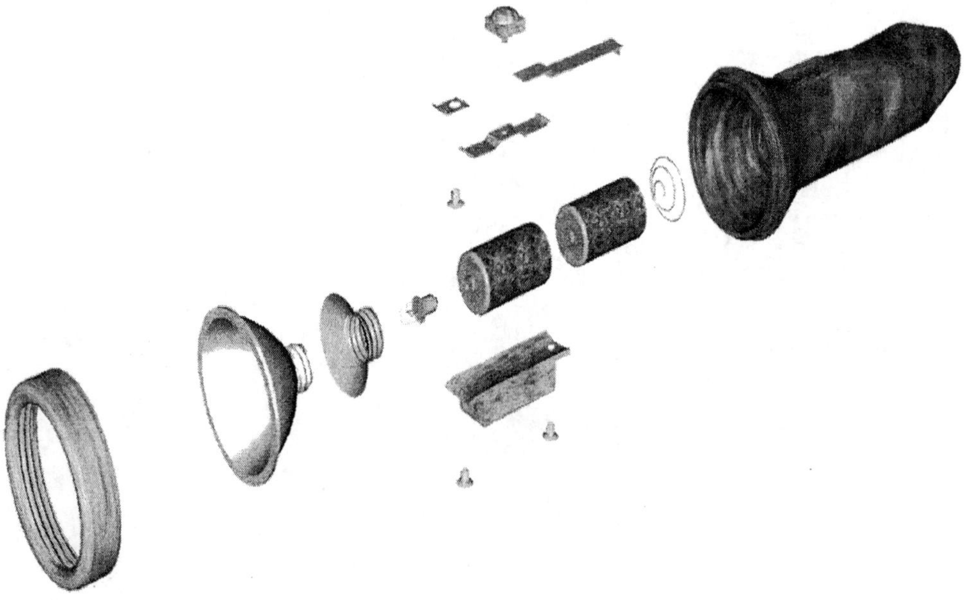

3-D Computer Models